BRASSEY'S
MODERN
FIGHTERS

BRASSEY'S
MODERN FIGHTERS

The Ultimate Guide to In-Flight Tactics, Technology, Weapons, and Equipment

MIKE SPICK

Brassey's
Washington, D.C.

First published in the United Kingdom by Pegasus Publishing Ltd.

Copyright © 2000 Pegasus Publishing Ltd.

ISBN 1-57488-247-3 (alk paper)

Printed in Singapore on acid-free paper that meets the American National Standards Institute Z39-48 Standard

Brassey's
22883 Quicksilver Drive
Dulles, Virginia 20166

First Edition

10 9 8 7 6 5 4 3 2 1

THE AUTHOR

Mike Spick is a leading commentator on military aviation with more than twenty-five books to his credit including *Modern Air Combat*, *Illustrated Guide to Modern Attack Aircraft*, *Fighters At War*, and *Classic Warplanes*.. He is currently a consultant to *AirForces Monthly* and a contributor to *Air International* and *Air Enthusiast*, as well as other respected international journals.

Editor: Ray Bonds
Creative Director: John Strange
Designer: Richard Carr
Color artists: Terry Hadler and Steve Seymour (via Bernard Thornton Artists); artwork of F-14 Tomcat, F-15 Eagle, F-16 Fighting Falcon, F/A-18 Hornet and MiG-31 Foxhound © Salamander Books Ltd.
Photo researcher: Tony Moore
Diagram artist: Michael Haywood
Color separator and printer: Acumen Overseas Pte Ltd, Singapore

CONTENTS

INTRODUCTION

Opposite: A two-seater F/A-18D
Hornet displays its power and
agility with a vertical climb.
Hornets will be in service with
many air arms around the world
for at least two more decades.

Modern fighters are breathtaking in their capabilities. Most can achieve speeds of more than one nautical mile (1.85km) every three seconds, and altitudes in excess of 9nm (16.67km). The fighter does not have to see its prey; it can detect and destroy it from far beyond visual distance, often using data from distant sources. It can perform barely believable gyrations. And if it detects a threat, its clever systems can not only automatically attempt to counter it they can warn its pilot of the best action to take. In the latest fighters, the pilot can talk to his aircraft, telling it what he wants it to do!

The modern fighter stands at the pinnacle of air combat technology. Less than 90 years separates the "stick and string" flying machines of 1914 from the wondrous weapons systems of today. For that is what the fighter has become: a weapons system.

Warfare has always been a matter of strategy and tactics. To take a simplistic view, strategy is aimed at winning the war; tactics at winning the battle, encounter, skirmish, etc. The best definition of tactics is the combination of fire and movement. In other words, bringing the available weaponry to bear to the greatest effect. Advances in weaponry inevitably bring about changes in the ways that battles are fought.

Cost always has been and always will be a factor in war, since it affects availability of weapons systems. The equation is: cheap and plentiful, or expensive and few. The mobility of the heavily armed and armored mounted knight of medieval times proved effective against foot soldiers. The balance was redressed by the long-ranged, fast-firing and accurate English longbow during the Hundred Years War between England and France, which mowed down the French knights before they could close with the enemy.

The longbow had just one fault. It was a highly specialized weapon, which required years of training to use effectively. This is why it was finally supplanted by the slow-firing, inaccurate and relatively short-ranged but cheap musket, which could be used by any conscript after minimal training. This is just one of the many recorded cases in which quantity overcame quality.

The breech-loading magazine rifle and the fast-firing machine gun not only replaced the musket, but made cavalry far too vulnerable. Yet the need for reconnaissance, and the equal need to deny reconnaissance to the enemy remained. The aeroplane, with its ability to turn the vertical flank of the enemy by over-flying him, provided the first. The corollary was to arm aeroplanes to deny the sky to enemy reconnaissance machines. Thus was born the fighter.

For decades the primary fighter weapon was the fixed, forward-firing machine gun or cannon. The variations were wide: from between one and 12 rifle caliber machine guns to up to four 30mm cannon. The problem then became to bring the guns to bear. High speed and rate of climb were needed to catch and force battle on an opponent; once this had been done, superior maneuverability was needed to get the enemy in the gunsight.

This situation pertained until the late 1950s, when the advent of the homing missile, able to track and follow its target, made combat possible at longer ranges than guns allowed. The dominant factor in air combat had always been the element of surprise; to attain this in the missile era, even earlier detection was needed. This was provided by radar and, to a lesser degree, by infra-red. The latter of course gave no indication of range, unlike radar. This gave birth to the true weapons system, in which detection and tracking played an equal part with the weaponry. The fighter became a platform, albeit an extremely mobile one, on which to carry detection and tracking systems, and weapons.

Over the years, two distinct patterns have emerged. The first is that beyond visual range (BVR) combat is the way to go. This combines superior detection systems, coupled with low observables, or stealth, to enable a first shot to be taken before the danger zone is entered. As this cannot be guaranteed, the second pattern is super-maneuverability, which gives the advantage if the fight closes to knife-range.

Which of the two endgames will predominate remains to be seen. Much depends on scenario (one versus one; few versus many, etc.), and the specific weaponry, avionics fit, and force multipliers of the respective opponents. But theory aside, combat is the ultimate arbiter.

Mike Spick

THE RIGHT STUFF

The *Right Stuff* was the name of a book and a film about test pilots and astronauts. Basically it posed the question: what was the factor which gave a man the uncritical willingness to face danger in test flying, which in the 1950s and 1960s was far from a good insurance risk, and also in the early space missions? The answer was that he had "The Right Stuff," a self-regard which convinced him, against all evidence to the contrary, that he had the ability and the nous to succeed where lesser mortals failed. The aircraft themselves have to be able to outfly and outfight the opposition. Whereas knights of old wore armor of plate, the modern knights of the air wear the invisible but magic armor of confidence in technology.

Fighter design is inevitably a compromise between different requirements. It is modified by two main factors: the nature of the perceived threat, and the air - the medium in which it operates.

Threats take many forms. The "worst case" scenario is the high-flying supersonic strike aircraft, carrying nuclear-tipped cruise missiles which penetrate at low altitudes. To intercept these, quick reaction and high performance are essential, backed by the ability to detect and destroy small and fast targets if the missiles are launched before the carrier aircraft can be intercepted. At the other end of the defensive scale is a purely tactical scenario, in which conventionally armed intruders have to be prevented from penetrating friendly air space.

The offensive scenario must also be considered. Air superiority must be gained in order to allow friendly attack or reconnaissance aircraft to carry out their missions unmolested. This is extremely difficult. Air defense is not simply a matter of opposing fighters; it is a cohesive entity of surface-to-air missiles (SAMS); anti-aircraft artillery (AAA), and ground-based radar and fighter control systems. Fighter missions, whether air superiority, combat air patrol, or strike escort, must operate in a hostile environment, unlike the defensive scenario, in which the environment is largely friendly, or at worst, neutral.

The medium in which fighter pilots operate is the air, the atmosphere of our planet. This is extremely variable. At sea level it is relatively dense, but as a rule of thumb air density decreases at slightly less than two percent per 1,000ft (305m) of altitude. This is both good news and bad news. The good news is that the dense air at sea level provides plenty of oxygen to allow the jet engine to reach maximum thrust. The bad news is that the denser the air, the greater the aerodynamic drag. To a degree, the one offsets the other.

The atmosphere itself is variable. For all practical purposes it is divided into the troposphere and the stratosphere; the boundary between them is called the tropopause. In the troposphere the air gets thinner and colder with altitude, and the speed of sound, referred to as Mach number by the West, and as the Bairstow number by Russia, gradually reduces. Above the tropopause, the air gets even thinner, but the temperature remains constant, as does the speed of sound. The height of the tropopause varies considerably with season and climate, but for purposes of calculation the ICAO Standard Atmosphere puts it at 36,090ft (11km), and it is here or slightly above that the optimum balance between thrust and drag is achieved, and maximum performance is reached.

Maneuverability is another matter. The most common form is turning ability, but possibly even more important is rate of roll or pitch and, to a lesser degree, yaw, which not only determine how rapidly direction can be changed, but confer pointability. This is known as transient maneuverability. For many years this was a purely aerodynamic function, but in recent times thrust vectoring has revolutionized transient maneuverability, and opened up a whole new area of the flight envelope.

Performance and maneuverability exist for two reasons only: to bring weaponry to bear on the enemy, and to evade his weaponry.

ABOVE: The Tornado F.3 was designed for long range, semi-autonomous defensive operations far out over the North Sea. Its engines were optimized for economy, at the expense of true fighter performance.

BELOW: External fuel tanks and external missile carriage both compromise stealth. The Joint Strike Fighter is to carry all fuel and non-stealthy weapons internally, as this computer-generated image shows.

RIGHT: Internal space for fuel is at a premium, as seen here in this cutaway of the F/A-18 Hornet. Care has to be taken that, in the event of battle damage, fuel spillage does not enter the engines and cause an uncontrollable fire.

PROPULSION AND PERFORMANCE

Discounting the fact that there is a steep rise at around Mach 1, drag increases in proportion to the square of the speed. Therefore doubling the speed quadruples the drag. To overcome this calls for four times the thrust. A given amount of aviation fuel contains a finite amount of energy, and it is the function of the fighter engine to produce the maximum thrust from this as economically as possible.

Maximum thrust is achieved by using afterburning, in which fuel is sprayed into the hot engine efflux and ignited. This is most uneconomical. As a fighter carries a limited amount of fuel, the effect of afterburning on combat endurance is potentially disastrous.

BELOW: Rafale carries three large external fuel tanks to extend range and endurance, but at a penalty in performance and weapons load. As they can also compromise stealth, their main value is for ferry flights over long distances.

For many years it has been standard procedure for fighters to carry additional fuel in external tanks. While this increases combat radius, and also the time that the fighter can spend in afterburner, it has certain disadvantages. Firstly it sterilizes hardpoints on which extra weaponry can be carried. Secondly the tanks create extra drag. Not only does this reduce performance, but generally half the extra fuel in the tanks is used in taking the other half to the point at which it becomes effective.

The alternative is to use in-flight refueling. This is fine provided that the vulnerable tankers can be kept sufficiently far back from the front lines as to be out of harm's way, but this in itself poses an operational restriction. Be this as it may, air refueling looks set to continue for the foreseeable future. At a tactical level, "buddy" refueling pods can be carried by fighters, but the general effect of this is to halve the number of effective weapon carriers available.

From a design point of view, adequate range for the baseline mission on internal fuel is desirable. The trend has been to pack fuel in wherever it will go, using fuselage, wing, and even fin tanks. There are limits, however. To pack fuel tanks into every spare space causes a measurable increase in combat vulnerability, while imposing a penalty in increased structural weight and plumbing. A classic example was the Russian MiG-25 Foxbat, optimized for high speed and high altitude interception. Even though its structure was mainly of nickel steel, its 14 tons of internal fuel restricted maneuverability loadings to a little over 2g at maximum take-off weight, and 5g with 50

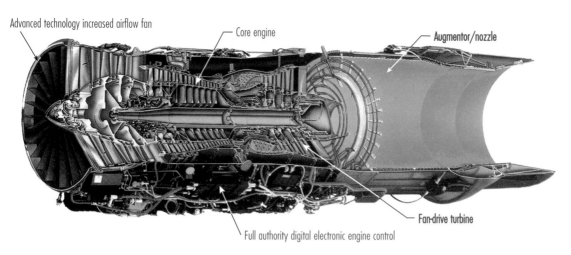

Advanced technology increased airflow fan

Core engine

Augmentor/nozzle

Fan-drive turbine

Full authority digital electronic engine control

LEFT: Cutaway of the Pratt & Whitney F100-PW-229 augmented turbofan, showing the two sizes of compressor. The difference in diameters is due to the bypass ratio. Both are driven by turbines aft of the flame cans, via coaxial spools.

percent fuel load. For a fully agile fighter, the optimum fuel fraction is about 0.30.

The search for greater range led to the axial flow turbojet being supplanted by the turbofan, in which the compressor, at the front end, overlapped the main body of the engine. The bypass air so created helped to cool the turbine section, parts of which operate above the melting point of the alloys used. This is most notable in airliner engines, in which economy is the greatest requirement; they have huge bypass ratios. But to achieve the required performance, fighter engines have a relatively low bypass ratio, to minimize frontal area among other reasons. So extreme was this that the forerunner of the F404 powerplant of the F-18 Hornet was at one point dubbed "the leaky turbojet"!

Some modern fighters are single-engined; others have twin-engines. There are various reasons for this. Back in the 1970s, two engines were regarded as having a wider margin of safety. This was of course more true of peacetime flying than in times of war; engine failure on a twin still left one operative with which to recover to base, whereas the combat record clearly shows that catastrophic engine damage tended to be equally deadly to both singles and twins.

Since then, engines have progressed considerably. They are simpler, have fewer parts, and have become far more reliable, to the point where losses to engine-related causes are minimal. Fighter loss rates per 100,000 hours have cumulatively almost halved for every decade since the 1960s, with the result that safety and survivability are no longer valid factors in the single versus twin controversy.

ABOVE RIGHT: Single-engined fighters like this F-16 have advantages in simplicity and affordability but maximum take-off weight is limited by the thrust available from a low bypass ratio turbofan.

RIGHT: Great capability means greater weight, which in turn requires two engines to attain the necessary thrust/weight ratio for the desired level of performance. Seen here is the Eurofighter Typhoon.

Fighter engines are designed to be simple, reliable, and compact, with a small frontal area. Given these constraints, the available thrust is limited to, as a ballpark figure, 20,000lb (9,070kg) in military power (that is, non-afterburning), and 30,000lb (13,600kg) with afterburning.

A modern fighter needs to be able to take off in a very short distance, to climb rapidly, to accelerate like a scalded cat in order to replace energy that has been bled off by hard maneuvering, and to be able to sustain turns at high g-loadings. It is also expected to be able to supercruise: to maintain level flight at speeds at Mach 1.2 or more on military power only. For all these it needs a thrust/weight ratio in excess of 1:1 in the fighter configuration. If the fighter is to be used in the ground to air role, to deliver bombs, this ratio may be relaxed, but for air combat it is close to essential.

The arithmetic is simple: if the best available engine has an output of 30,000lb (13,600kg), adequate performance in the fighter configuration will be achieved only if take-off weight is something less than this figure. Once it is significantly exceeded, only a twin-engined layout will suffice.

Once this point is reached, considerable weight increases follow. Two engines demand a larger structure weight to carry them, which in turn makes for a larger airframe. More fuel and oil capacity is needed, with its associated plumbing, and more computer capacity is required to monitor them. Sturdier landing gears are needed to cope with the increased weight. Yet with all these increases in size and weight, maneuverability must not be allowed to suffer. A greater wing area is needed in order to keep wing loading within acceptable limits. The end result is a very much larger and heavier, and less affordable, airplane.

Operationally there are other problems. Engine maintenance is doubled. One could cynically argue that there is twice as much to go wrong! The conclusion is that single-engined fighters are preferable unless the warload/range combination is such that two engines are needed to provide adequate power for the projected missions.

Supercruise appears to be a relatively new concept, but in fact the underlying principles are as old as air warfare. The only difference is that whereas in the past opposing aircraft could stay at full throttle for extended periods, the advent of afterburning dramatically reduced the time that fighters could stay at supersonic speeds. Supercruise, which gives speeds of Mach 1.4 or more in military power, extends the time available in supersonic flight by an order of magnitude.

ABOVE: Vortices, made visible by condensation, stream from the leading edge extensions of an F-16 as it hurtles skywards. High acceleration and rate of climb is essential in a modern fighter.

The critical point arises from the weapons load that the fighter is designed to carry. Medium range air-to-air missiles (AAMs) are fairly large, and to carry a significant number a large airframe is needed. It is of course quite possible to hang several AAMs on a small aircraft, but the general effect is like hanging an anchor on the airplane: performance and agility suffer. The point is soon reached where only a larger airframe will suffice.

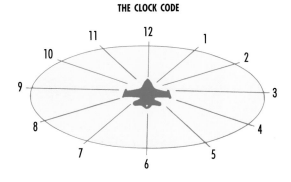

THE CLOCK CODE

RIGHT: The clock code is used for orientation against other aircraft. The fighter is in the center of an imaginary clock: 12 o'clock is ahead; six o'clock is astern. A contact to the right and above would be called as "3 o'clock high."

Above: Twin engines were adopted for the Dassault Rafale for the same reasons as for the Typhoon. This was an advantage for Rafale M, seen approaching the deck during carrier compatibility trials, as this configuration gives added safety for carrier operations.

Right: Speed considerably shrinks the lethal engagement envelope of surface-to-air missiles. Here it can be seen that the faster aircraft is at risk for less time. Unlike afterburning, supercruise can be sustained for long periods.

SAM ENVELOPES WITH TARGETS AT DIFFERENT SPEEDS

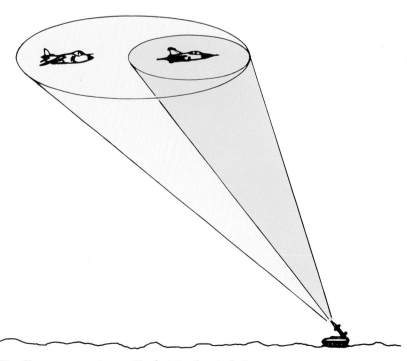

The phases of battle are: detection, closure, engagement, maneuver (if necessary), and disengagement. Historically, fighters have needed to catch their opponents and force battle upon them. For this, they needed a simple speed advantage. Against conventional opponents, supercruise provides this in spades, reducing the time taken to close to weapons range, and giving the ability to run down a fleeing target. The closure phase will therefore be much shorter and engagement more certain. Disengagement will equally benefit. Sheer speed will make the supercruise fighter very difficult to intercept. Even if an adversary manages to get into the preferred six o'clock position, its missile envelope will be badly crimped in, while the use of military power by the supercruise fighter will keep its heat signature relatively low. Its opponent will of course have to use afterburner to stay in position, and the chances are that it will run out of fuel in a matter of minutes.

Finally, supercruise will shrink the lethal envelope of surface-to-air missiles quite considerably. The difference between Mach 1.4 and Mach 0.9 will reduce the time available for detection, tracking, and missile launch by at least one third, while providing a relatively poor heat target.

AERODYNAMICS, STRUCTURES, AND AGILITY

The raison d'être of a fighter has always been to catch and destroy enemy aircraft. For this the defender has needed performance, speed and rate of climb, to catch it and force it to battle. This hopefully is enough to take the opponent by surprise, giving the fighter an easy, non-maneuvering target. When all else fails, the target must be outmaneuvered and shot down. Maneuverability is also a defensive attribute, enabling the fighter to stay outside the missile/gun envelope of its opponent.

There are two types of maneuverability: transient and turn. Transients are the ability to change from one flight mode to another in three planes, pitch, roll, and yaw. Pitch is the ability to change direction vertically to the axis of the aircraft and to climb or dive if in level flight, which is induced by movement of the horizontal tail surfaces. When a conventional fighter goes supersonic, the center of lift moves aft, and the tail surfaces have to provide a download to preserve stability. The result of this is excessive stability at supersonic speeds, with a consequent loss of maneuverability. Modern fighters are built with relaxed stability to avoid this, using computerized fly-by-wire or fly-by-light to give the correct control inputs.

Yaw is of little interest to us, but roll, induced by ailerons on the wings, or in times past by spoilers which dumped lift on one side or the other, is very important. Rate of roll determines how quickly the fighter can commence a hard turn, while rate of acceleration into the roll is even more important.

Turning ability is largely determined by the amount of lift a wing can generate at a given speed/altitude combination. The factors involved are the coefficient of lift, which these days varies little between one fighter and another; wing loading, the ratio of weight to wing area; and the aspect ratio, which is the square of the span to the wing area. Of these, wing loading is the most important. As a general rule, the fighter with the lowest wing loading will out-turn its opponent.

Turning ability is measured in two ways: radius of turn, measured in feet, and rate of turn, measured in degrees per second. Turning is also often referred to in terms of multiples of the force of gravity, or g. In level flight, lift balances the weight of the aircraft. In a hard turn, more lift is needed to counter the centrifugal force involved. To obtain this, the angle of attack of the wing, or alpha, as it is generally known, must be increased, which in turn causes increased drag, which needs more thrust to counter it.

FIGHTER MANEUVERABILITY STANDARDS: ROLL, PITCH, AND YAW

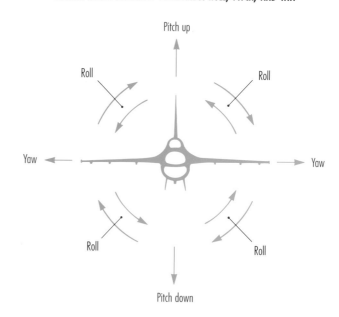

Pitch up

Roll

Roll

Yaw

Yaw

Roll

Roll

Pitch down

There are two forms of turn: sustained and instantaneous. Sustained turn is precisely what its name implies: it is the radius and rate of turn which can be held indefinitely for the speed/altitude combination, where thrust exactly equals drag. Instantaneous turn is the hardest possible turn at the speed/altitude combination, but in which drag exceeds thrust, which bleeds off speed, often at an alarming rate. When this occurs, only two possibilities exist. Either the turn must be relaxed to the point where sustained turn performance is reached, or speed will continue to be bled off until the fighter reaches the edge of the 1g envelope, where all turning capability is lost. The latter leaves the fighter flying straight and level on the edge of a stall, which in combat maximizes vulnerability.

ABOVE: To change direction, a fighter must first pitch up or down, yaw sideways, or roll. Transient maneuvers such as these, are the preliminaries to changes in flight path.

LEFT: The Lockheed Martin F-16 set new standards for agility when it first entered service. Even today it is hard to beat by any really significant margin.

BELOW: Heavily laden with drop tanks, air-to-surface weapons and air-to-air missiles, Rafale retains a remarkable amount of agility.

WING AND THRUST LOADINGS

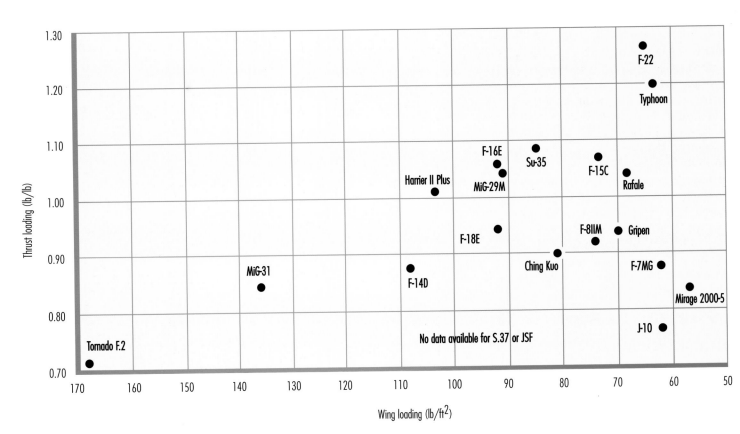

Wing loading (lb/ft²) — horizontal axis: 170, 160, 150, 140, 130, 120, 110, 100, 90, 80, 70, 60, 50

Thrust loading (lb/lb) — vertical axis: 0.70, 0.80, 0.90, 1.00, 1.10, 1.20, 1.30

F-22, Typhoon, F-16E, Su-35, MiG-29M, Harrier II Plus, F-15C, Rafale, F-18E, F-8IIM, Gripen, Ching Kuo, MiG-31, F-14D, F-7MG, Mirage 2000-5, J-10, Tornado F.2

No data available for S.37 or JSF

ABOVE: The agility needed for close combat calls for low wing loading coupled with high thrust loading. On this basis, Typhoon and the F-22 Raptor should be outstanding.

RIGHT TOP: Corner velocity is where radius of turn and rate of turn maximise. It varies with aircraft type. Combat should be begun at a rather higher speed, to allow for energy bleed-off.

RIGHT BOTTOM: Different wing shapes for different purposes. Swept wings delay the drag rise; high aspect ratio is good for sustained turn and economical cruising; low aspect ratio is best for supersonic flight; deltas give a large lifting area and low loading; variable sweep combines high and low aspect ratio; while forward sweep has many benefits.

Another way of looking at fighter performance and agility was the concept of energy maneuverability, formulated by USAF Colonel John Boyd in the late 1960s. There are two forms of energy: positional, in the form of altitude which can be converted into speed by diving, and kinetic, which is essentially a combination of the mass and velocity of the fighter. Basically, this concept stated that, since all maneuvers consume energy, that energy must be carefully managed in combat to prevent it from being squandered.

With all fighters, there is a combination of lift and limited speed/altitude where the smallest radius of turn and highest rate of turn are reached. This is called corner velocity (V_{corn}), which is generally about 300-400kt (556-741km/h). Above V_{corn}, sustained turn rate in degrees per second falls away fairly slowly but radius of turn widens rapidly. Below V_{corn}, both radius and rate of turn fall away quite rapidly. Ideally, maneuver combat should be entered well above V_{corn}, which allows a fair amount of energy to be bled off in very hard maneuvers. But once V_{corn} is reached, energy must be carefully husbanded if the fighter is not to become, in the time-honored phrase, "out of altitude, energy, and ideas!"

Way back in the mid-1970s, a revolutionary fighter appeared. This was the General

Dynamics F-16, which could sustain a turn of no less than 9g, albeit over a small portion of the envelope. Prior to this, most fighters were stressed for between 7.33 and 8g maneuvers, with a one-third safety margin. While they could generally exceed these figures in instantaneous turn, few could so much as reach them in a sustained turn. The F-16 established a maneuver performance plateau which even today, over 20 years later, is still hard to exceed by a significant margin.

A 9g sustained turn is certainly impressive, but what is its value in combat? At 500kt (927km/h) it gives a turn radius which is about 720ft (219m) smaller than is possible from 7g at the same speed, plus a couple of degrees per second angular advantage. The down side is that at 9g the pilot is strained to his physical limit, with blackout threatening, while 9g is also pushing missile launch limits rather too hard. However, in a defensive situation it compounds the difficulties of an attacker seeking a firing solution. Or does it? An old pal who flew F-5Es for the USAF's Aggressors dissimilar air combat training unit once told me that, faced with an F-16 in a series of sustained 9g defensive turns, "I just flew around him a while and let him get dizzy!"

Sustained turning ability is nice to have, but in combat its value is questionable. One school of thought maintains that the only time sustained turn is reached is when the fighter passes through it while using instantaneous (maximum) turn rates.

Conventional fighter maneuverability is based on the lift generated by the wings. Wing design is therefore very much of a compromise. A long wingspan maximizes lift without needing excessive alpha, and coupled with a low wing loading is best for small radius turns. On the other hand, a long wingspan slows the rate of roll, which in turn increases the time a fighter needs to establish itself in the turn. The compromise is therefore primarily between aspect ratio and wing loading, roll rate and lift. Attempts to overcome this were made with variable sweep wings, with a wide variation in aspect ratio dependent upon the sweep angle, but the main problem was that wing area was limited, and wing loading consequently high. A far better solution was to use computerized

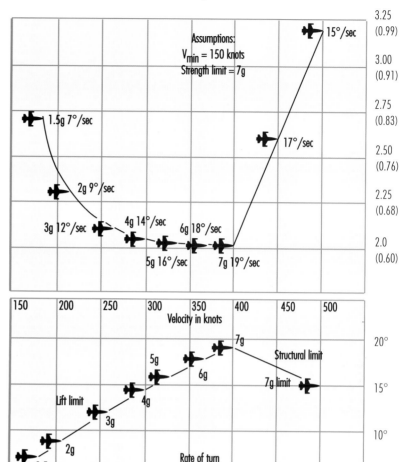

RADIUS OF TURN/RATE OF TURN

Assumptions:
V_{min} = 150 knots
Strength limit = 7g

Radius of turn in 000ft (m)

Velocity in knots

WING TYPES

Aspect ratio: is the ratio of the square of the span to the area.

High aspect ratio gives the most lift for the least drag at subsonic speeds. It gives high ceilings, economical cruising, and good sustained turn.

Low aspect ratio gives good supersonic performance and acceleration, rapid rate of roll, at the cost of excess speed loss during hard turning.

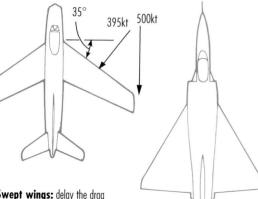

Swept wings: delay the drag rise. The velocity of the airflow normal to the leading edge is effectively reduced by a factor of the cosine of the sweep angle, as seen here.

Delta wings: low aspect ratio, large lifting area for low wing loading, simple construction, and depth for fuel tanks.

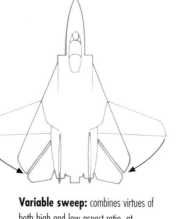

Variable sweep: combines virtues of both high and low aspect ratio, at the expense of structural complexity and high wing loading.

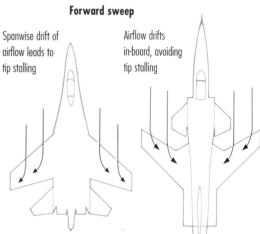

Forward sweep

Spanwise drift of airflow leads to tip stalling

Airflow drifts in-board, avoiding tip stalling

Variable camber: both leading and trailing edges are pivoted and can hinge up or down.

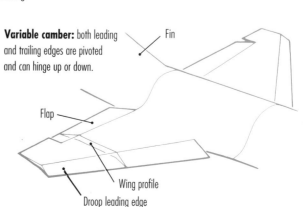

Fin

Flap

Wing profile

Droop leading edge

LEFT: The problems of the tailless delta were largely overcome using relaxed stability and fly-by-wire (and later fly-by-light) coupled with canard foreplanes. Dassault used this format to good effect with Rafale.

BELOW LEFT: With fly-by-wire, pilot input on the controls is translated into a demand which is then processed by computer and sent to the hydraulic actuators which move the flight control surfaces. This results in a large weight saving over a conventional hydraulic system.

FLY-BY-WIRE FLIGHT CONTROL SYSTEM

Electric signalling (position demand)

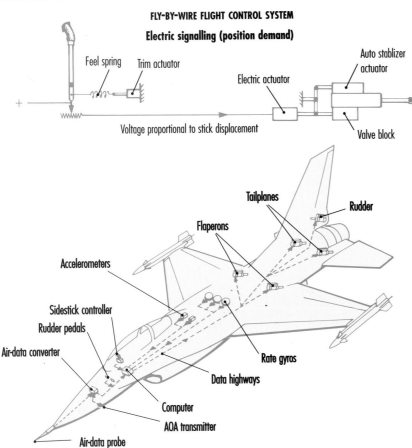

Feel spring   Trim actuator

Electric actuator

Auto stablizer actuator

Voltage proportional to stick displacement

Valve block

Tailplanes   Rudder

Flaperons

Accelerometers

Sidestick controller

Rudder pedals

Air-data converter

Rate gyros

Data highways

Computer

AOA transmitter

Air-data probe

variable camber to optimize lift according to the flight regime.

The tail-less delta configuration (in, for example, the F-102 and F-106 fighters of the 1960s) offered many advantages. It combined a sharp leading edge sweep angle with a large lifting area, and even within the constraints of a low thickness/chord ratio, essential for supersonic flight, it had sufficient volume to house both the main gears and capacious fuel tanks. Its main drawback was a very low aspect ratio. Maximum lift was created at high angles of attack; this, combined with the lack of horizontal tail surfaces, meant that take-off and landing runs were long, and high drag in hard turning caused speed to bleed off at an excessive rate.

The worst problems of the tail-less delta were cured with the advent of relaxed stability and fly-by-wire, but an even better solution was to combine it with canard foreplanes. Whereas at supersonic speeds conventional tail surfaces had to provide downloads to compensate for the aft movement of the center of lift, foreplanes added to the overall lift, as well as improving supersonic maneuverability.

WEAPONRY

GUNS AND GUNNERY

For nearly half a century, the primary fighter weapon was the gun. It was then supplanted by homing missiles, but the shortcomings of these were quickly revealed in combat and the gun was reinstated. Modern fighters still carry guns, although for air combat they are now regarded as weapons of the last resort, to be used when all else has failed. They can, however, be used effectively in the secondary role of strafing ground targets. Just one quibble: as guns on "real" fighters, as opposed to a few armed trainers, have a caliber of more than 15mm, they are technically cannon, and fire shells rather than bullets. But as the technique of air-to-air firing is known as gunnery rather than cannonry, let us stay with guns!

Gun targets in close combat are generally fleeting, and unlike missiles, which require a certain amount of "switchology," the gun has

LEFT: Although short-ranged, the gun is an instantly available weapon for split-second shooting opportunities in close air combat. It can also be, as seen here, a valuable weapon with which to attack ground targets.

BELOW: A MiG-29 Fulcrum, crossing at 90 degrees and 300kt, could be hit a maximum of nine times by an M61 cannon firing at 6,000 rounds per minute, but only if aim was along its axis. Even a slight angular deviation would reduce the number of hits dramatically, reducing the probability of a kill.

the advantage of being instantly available. Like all else, aircraft gun design is a compromise. The twin needs are to score hits, and to provide adequate destructive power.

To hit an evading target, the need is to put a lot of shells in the air very quickly - i.e., a high rate of fire. Let us assume a target with an angular crossing speed of 200kt (371km/h), which subtends an effective length of 30ft (9.14m). Given a true aim down its length, a rate of fire of 800 rounds/min could expect to score just one hit, with a less than 20 percent chance of a second. Moving up to 1,200r/m, the expectation would be one hit, with a 77 percent chance of a second. Taking the ultimate in rates of fire, 6,000r/m would give nearly nine hits.

A high rate of fire is given by one of two methods. The first is the revolving chamber feed gun as pioneered by the 30mm German Mauser MG 213 in 1945, and later adopted for the British Aden, the French DEFA, the American Colt M39, and the Russian Nudelmann-Richter NR-30. All were 30mm with the exception of the M39, which was 20mm. Cyclic rate of fire was around 20 rounds per second. Where they differed tremendously was in the shell weight. The American 20mm shell weighed a mere 3.66 ounces (104 grammes), whereas the Aden and DEFA shells were virtually identical at 7.99 ounces (227 grammes), while the Russian shell weighed in at a massive 14.47 ounces (411 grammes). The latter had almost double the propellant charge of the Aden and DEFA for a comparable muzzle velocity. Naturally,

GUNFIRE ALONG AIRCRAFT AXIS

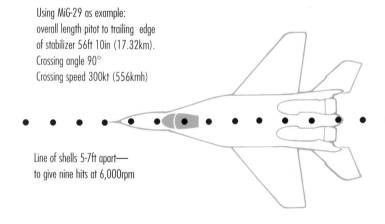

Using MiG-29 as example:
overall length pitot to trailing edge of stabilizer 56ft 10in (17.32km).
Crossing angle 90°
Crossing speed 300kt (556kmh)

Line of shells 5-7ft apart—
to give nine hits at 6,000rpm

LEFT: A General Electric M-61A cannon from a Belgian F-16, seen here stripped for servicing. The six rotating barrels, breech, and feed mechanism are clearly visible. The large cylindrical object to the right is the ammunition tank.

the heavier shells have the greater destructive capacity, but as a rule the heavier the shell, the slower the rate of fire.

The second method of obtaining a high rate of fire was the rotating multi-barreled gun working on the Gatling principle. The first of these, and in many ways the best, was the General Electric M61 Vulcan which is still in use. Once it has worked up to maximum rate, which takes 1/4 second, it can churn out 6,000 shells per minute. Its main failing is the standard 20mm shell, which not only lacks punch, but has poor ballistic qualities. Modern Russian aircraft guns work on the same principle, but are twin-barreled, giving a slower rate of fire, using 23mm or 30mm ammunition.

Modern combat aircraft, designed to withstand incredible stresses and strains in the air, can also survive a high level of battle damage from guns. However, the consensus of opinion is that 20mm shells are demonstrably too light. As caliber and shell weights increase, a greater propellant charge is needed to produce sufficient muzzle velocity, and thus adequate ballistic qualities, and also a larger and heavier gun. The current ultimate in aircraft guns is the GAU-8/A fitted to the USAF's A-10 Warthog ground attack machine, which in effective range, hitting power, and muzzle velocity, outperforms all others. But as the gun is the size of a Volkswagen Beetle automobile, it is far too large and heavy to be housed in an

BELOW: Whereas the standard US aircraft gun is 20mm, the latest Russian fighters carry the 30mm GSh-301, the muzzle of which is just visible above the wing root of this Su-27 Flanker. Much less ammunition is carried, but the 30mm shell is more destructive than the American 20mm.

agile fighter, and is restricted to tank busting. The compromise between rate of fire and destructive power has seen the emergence of a range of fighter guns of around 25mm or 27mm caliber.

One of the most intractable features of air-to-air gunnery is aiming. Once fired, a shell leaves the barrel at the muzzle velocity of the weapon, typically about 3,000ft/sec (914m/sec), with added kinetic energy imparted by the speed of the fighter. It immediately meets with aerodynamic drag (which varies with altitude and climatic conditions), and starts to slow down. Its kinetic energy is more or less expended by the time it has traveled 2nm (3.7km), during which time it will have fallen 100ft (30m) or more, at an ever-increasing rate, due to gravity drop.

If the fighter was maneuvering when it fired, two other factors would come into play. A side-force would be imparted to the projectile, giving what is known as bullet trail, while the relative wind (the slipstream passing the fighter) would be off-boresight (boresight being the direction that the longitundinal axis of the fighter is pointing). Finally, since the target would be moving, if not actually maneuvering, this would mean that, in order to score hits, the projectiles would have to be aimed ahead of the target. This is known as deflection shooting.

Deflection shooting with a fixed sight calls for extremely fine judgement of range and relative speed, and very few fighter pilots have ever mastered it. The traditional method was always to get in so close that it was nearly impossible to miss, but in the present age this is not feasible. Even if it could be done, bouncing around in the jetwash of a modern fighter would prevent accurate aiming, while debris from an opponent at close range could well be ingested by one's own engine, which would be counterproductive. The task of accurate aiming being far beyond the abilities of the average pilot, therefore, it has largely been handed over to the clever electrons. Fighter speed, altitude, attitude, and g-loadings are fed into the fire control computer, where they are combined with target data provided by tracking, and accurate range provided by radar. The sight, an illuminated dot usually surrounded by a circle, known as a disturbed reticle system, then wanders across the head-up display (HUD) until it settles on the correct solution, and the pilot opens fire.

Guaranteed hits? Not exactly! Guns are inherently inaccurate, which causes bullet dispersal. In part, this is due to vibration and, with rotary cannon, the effect of barrel rotation, in part due to manufacturing tolerances and barrel wear. This is typically between

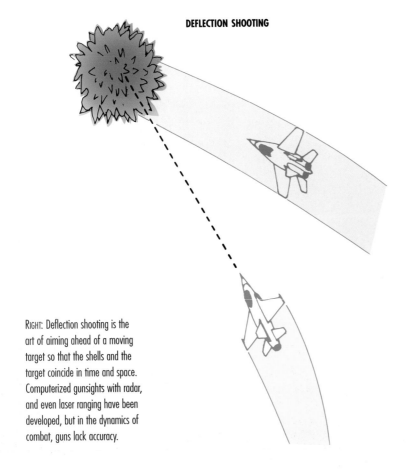

DEFLECTION SHOOTING

RIGHT: Deflection shooting is the art of aiming ahead of a moving target so that the shells and the target coincide in time and space. Computerized gunsights with radar, and even laser ranging have been developed, but in the dynamics of combat, guns lack accuracy.

MISSILE HOMING METHODS

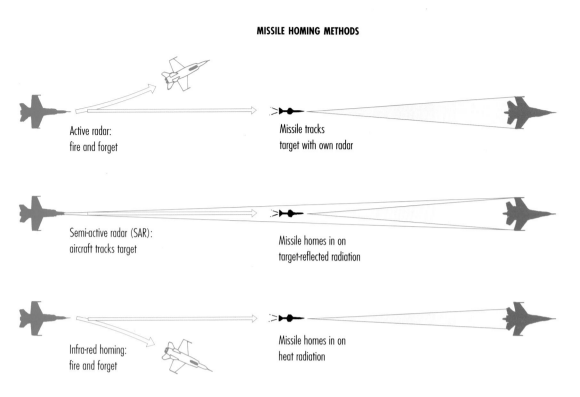

Active radar:
fire and forget

Missile tracks
target with own radar

Semi-active radar (SAR):
aircraft tracks target

Missile homes in on
target-reflected radiation

Infra-red homing:
fire and forget

Missile homes in on
heat radiation

LEFT: Three types of homing are in widespread use: active radar, which at medium ranges needs mid-course guidance from the launching fighter before it can close the distance sufficiently to start homing; semi-active radar, which needs the launching fighter to illuminate the target for its entire time of flight; and infra-red, a launch and leave weapon used for close encounters.

four and six mils, a mil being the angle subtended by an object 12in (305mm) long at a distance of 1,000ft (305m). For the record, one degree of angle subtends 171/2 mils. Given a typical firing range of 2,000ft (610m), a target with a wingspan of 35ft (10.67m) subtends an angle of one degree, giving a high probability of scoring hits.

MISSILES

Missiles able to home unerringly on their targets promised to revolutionize air warfare when they were first introduced in the 1950s, not least by eliminating maneuver combat. But they flattered only to deceive. After four decades in service, they are still unbeatable, and are rapidly becoming too expensive to use.

Initially there were five different missile guidance systems: wire-guided, beam riding, active radar homing (AR), semi-active radar homing (SARH), and infra-red (IR). The first two were quickly found to be impracticable and abandoned. Initially, so was active radar, although in recent times this has made a comeback. This left just SARH and IR, which are still in large-scale service.

To use a SARH missile, the fighter acquired the target on radar, tracked it, then activated the missile, which homed on the reflected radar energy. Consequently the fighter was forced to illuminate the target throughout the entire time of flight of the missile. This had three disadvantages: the earlier fighter radars became blind to all else while providing radar guidance, and the fighter itself was forced to continue to close on the target. Both made it

vulnerable to counterattack, especially as seeker sensitivity against a target with a small radar signature was limited, reducing the effective launch range. Finally, the radar emissions could alert the target that it was under attack, enabling it to evade or deploy countermeasures.

By contrast, IR-homing missiles, which guide on the heat source provided by the target, are fire-and-forget weapons. Once they are launched, the parent fighter is free to disengage, take evasive action, or seek another target.

BELOW: A Dassault Mirage 2000C test fires a Matra Mica (*Missile intermédiat de combat aerien*). Mica is unusual in the West in having both active radar and infra-red homing variants, although this has long been standard Russian practice for all except the smallest weapons.

FIGHTER RADARS

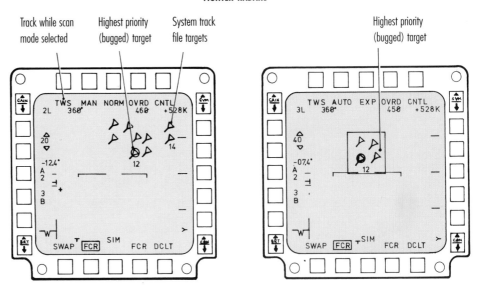

Track while scan mode selected

Highest priority (bugged) target

System track file targets

Highest priority (bugged) target

The Westinghouse APG-68 as fitted to the F-16C is a fairly typical multi-mode radar. Its air combat modes include the following:

Velocity search, which can detect closing, front aspect targets, at a range of up to 160nm, although they would need to have a large radar cross-section.

Track-while-scan, which continues to search while tracking up to ten targets. It can also assign target priorities (i.e. the greatest threat).

Downlook/uplook. Downlook uses Doppler filtering to screen out ground returns and detect targets at low level. Uplook gives one third better detection capability.

Dogfight is a whole raft of close combat modes which acquire targets automatically. Initially the radar scans a field of 30 x 20deg and locks on to the first target within 10nm. Boresight is a narrow pencil beam, while slewable is a vertical scan for picking up maneuvering targets.

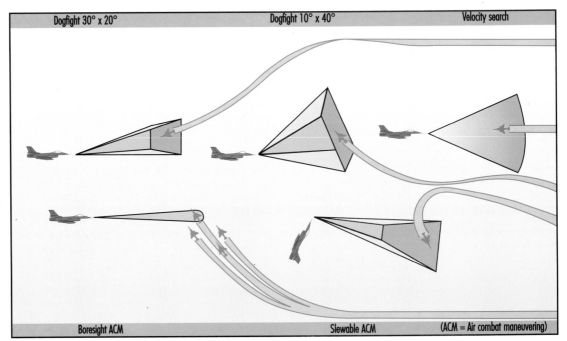

Dogfight 30° x 20°   Dogfight 10° x 40°   Velocity search

Boresight ACM   Slewable ACM   (ACM = Air combat maneuvering)

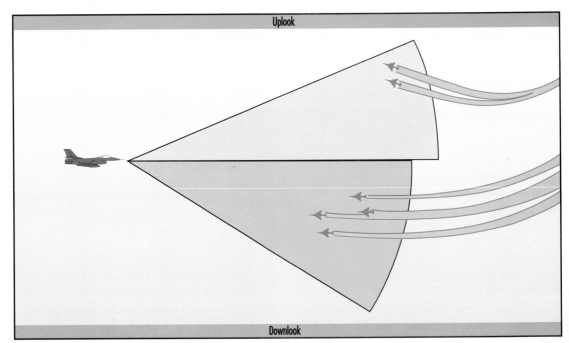

Uplook

Downlook

As a general rule, SARH missiles are medium range weapons, while IR-homers are reserved for visual distance encounters. There are several reasons for this. Radar-homing is inherently less accurate than infrared, and therefore it relies more on a near miss, and detonation by a proximity fuze, than does a heat-homer. The SARH missile therefore needs a large diameter warhead to give a high kill probability. A large diameter warhead means a large diameter body and a bigger rocket motor. Extending the length of this gives more rocket fuel and increased range. Operationally, radar can not only detect a target from far beyond visual range, but can keep the missile centered on that target. It can also see through weather that would blind a heat-seeker, and so remains effective in adverse conditions. Finally, it is far better in a head-on engagement than a heat-seeker.

The first IR-homers were limited to attacks from astern, where their seekers could "see" the hot engine efflux and tailpipe of the target. Since then, they have been given an all-aspect capability, where they can detect hot spots caused by kinetic heating anywhere on the target's airframe. Ability to do this varies, and IR-homers are still most effective when launched from astern. It is of course perfectly possible to produce medium range missiles with IR-homing, but this course is fraught with danger. In a confused situation, a heat missile might switch targets to a friendly aircraft with no possibility of recall. With a radar missile, this should not happen.

The drawbacks of SARH-homing have led to a greater accent on active radar (AR) in recent years. The initial problem was that power output, and therefore acquisition range, were restricted by the diameter of the seeker antenna. This has been largely overcome by using inertial mid-course guidance, which can be updated in flight. This also allows a rapid sequence of launches at multiple targets, using coded signals for each one.

The AR missile is launched towards a point in space where the fire control system predicts the target will be, in the time calculated for the missile to get there. The fighter's radar continues to track, sending data for course changes as necessary. When the missile nears the specified area, typically within 10nm (18.5km), the active radar seeker cuts in, acquires the target, and completes the terminal homing phase autonomously.

The first air interception missile to use this system was the AIM-54 Phoenix, a massive weapon 15in (381mm) in diameter and 13ft 2in (4.01m) long. Weighing 977lb (443kg), it needed a specialized aircraft (the US Navy's

F-14 Tomcat) to carry and launch it. Since then, miniaturization has reduced the dimensions of the new generation to the point where they closely match the size of the SARH weapons that they are to replace, and even quite small fighters can carry them, provided only that they have an adequate radar and fire control system.

It has been suggested that, as fighters are so dependent on radar, the ideal air weapon should be a small and simple missile that homed on radar emissions. Anti-radiation missiles already exist for attacking ground targets in this way,

ABOVE: A Magic 550 Mk2 curves hard across the bows of the launching Rafale as it starts to track a distant target. Modern missiles have become increasingly agile.

BELOW: An AIM-7 Sparrow launched by an F-16C streaks towards its target some 12 miles (19km) away. This was the first step towards giving the Viper BVR capability.

ABOVE: The Sidewinder is the most successful air-to-air missile of all, and over the years has been given extended range, more agility, all-aspect capability, and a wider look angle for the seeker. This is the latest variant, the AIM-9X.

and air combat trials, using a theoretical weapon, showed great advantages.

Alas, it was not to be. The complications of fighter radar operation have ensured that even if such a weapon could be designed and built, it would be very large, terribly heavy, and prohibitively expensive. It was the old story: you can always win the war with the weapon you haven't got!

AAM design poses extreme problems, one of which is obtaining the right combination of lightness and strength. They have to be able to withstand loadings of 50g or more on launch, or while maneuvering. The seeker

head cover has to be pretty special. Under normal conditions, it has to withstand scorching sun, before being exposed to rain and hail at supersonic speeds. It is exposed to temperatures of –56 degrees Celsius in the stratosphere, then brought down to lower levels very rapidly. After launch, it has possibly to endure prolonged kinetic heating caused by speeds of Mach 3-4. And all this without losing its dielectric qualities, its transparency, or its shape, all of which could be disastrous.

Most AAMs are powered by rocket motors, which accelerate them to maximum velocity in a matter of seconds. Despite what the brochures say, this is extremely variable; the dense air at low level can reduce it by a third or more. With the rocket all-burnt, it coasts on, the speed gradually decaying. Like an airplane, maneuverability reduces with velocity until, at the end of its run, the missile no longer has enough energy to follow its target.

AAM interception uses a process known as proportional navigation. When the seeker head acquires a target, it starts to steer, not directly at it, but at a point ahead of it, thereby cutting the corner with a collision course interception. All seekers have a "look angle": the angle off-boresight at which they can acquire and track a target. In the early days this was a mere few degrees, but has gradually been expanded out to 40 or even 60 degrees, with 90 degrees or more promised for the future. While this makes attacking rather easier for the fighter pilot, defending against this new breed of missile is far more difficult.

Missiles are usually controlled by means of small movable aerodynamic surfaces. Jet

RIGHT: A mix of Russian AAMs seen beneath the wing of a MiG-29. Outboard are R-73 dogfight weapons, NATO reporting name Archer, which at the time of writing had the greatest off-boresight capability of any missile in the world.

FIGHTER HOTSPOTS

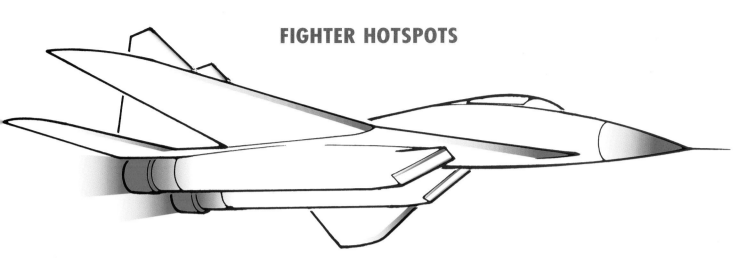

deflection is another possibility, either by movable vanes in the efflux, or by squibs, but once the motor is burnt out all vestige of control is lost.

In times past, hard maneuver by a target fighter was enough to defeat a missile, provided the pilot knew that it was on its way. This could take the fighter outside the look angle, or at worst make the missile seeker head bump its gimbal limits. Then, as lethality of missiles increased, a variety of countermeasures were used, electronic and infra-red. But the best counter of all, albeit the most difficult to execute, is to stay out of the missile envelope. To summarize, the modern air-to-air missile is an extremely advanced piece of kit, but it is not yet infallible.

STEALTH TECHNOLOGY

Stealth, or to give it its more appropriate title, low observability, is as old as war itself; its advantages are the attainment of surprise and tactical position. For centuries it was limited to two basic forms: covert approaches, often under cover of darkness, and carefully laid ambushes. It is therefore hardly surprising that, given its provenance, stealth should become a recognized part of air warfare.

What is surprising is that it began so early. Experiments to reduce the visual signature of airplanes took place in Austria in 1912, by replacing the fabric wing and fuselage coverings with transparent material. Of course, nothing could be done to hide the engine, the pilot and the guns, while the reflection of the sun was counterproductive. Finally, the trans-

parent covering was far heavier than doped fabric, thus reducing performance.

Although the experiment was repeated by Germany's Luftstreitkräfte on the Western Front in 1916, with Fokker Eindeckers, it was not pursued. This was perhaps just as well; in the multi-fighter dogfights which took place during the final two years of the war, identification and performance were of far more value than invisibility. The concept was sound, but its time had not yet come.

More than half a century passed before stealth even started to reach maturity, and by this time the threat had grown. Radar had expanded the visual detection distance of the Great War period by a factor of more than 20, although this was partially offset by increased aircraft speeds, which held the reaction time available to the defenses down to a factor of three or less.

Ever since its inception, radar has been the most important means of detection in the air defense system. Not only can it pierce darkness and adverse weather, whatever its shortcomings it is very good at providing accurate ranges, altitudes, and speeds at great distances.

Electronic countermeasures are nice to have for jamming radars, but they are not enough in the face of sophisticated modern systems. At the very least, jamming warns the radar operators that something hostile is out there, which alerts the defenses. What is needed is to elude the searching radar emissions altogether, and only stealth can provide this.

It is of course just as impossible to build a fighter that is invisible to radar as it is to produce one that is invisible to the human eye,

ABOVE: Infra-red seekers home on hot spots on the target. The plume from the engine exhausts is the largest, followed by the nozzles. From other aspects, aerodynamic heating maximizes on the nose, the intake lips, and the leading edges of wings, fins and tail.

RIGHT: The Sukhoi Su-27SK Flanker is an aerodynamically excellent fighter, but has no readily apparent stealth features. It is large, but surprisingly size has only a marginal impact on RCS. Shaping is far more important.

BELOW: The lines of the Lockheed F-22 Raptor are carefully angled to deflect radar emissions, while great care has been taken to avoid surface discontinuities. Where these are unavoidable, such as the front of the canopy, a serrated edge has been used for deflection.

but much can be done to minimize the probability of detection. Radar sends out electronic impulses, which attenuate rapidly with distance. When these strike a solid object, such as an airplane, they bounce off. Depending on the angle of the surface at which they hit, much of the radar energy is deflected off at an angle and lost, but inevitably a little is reflected back towards the sender where, given a sufficiently sensitive receiver, it can be detected.

Surprisingly, the physical size of the airplane under surveillance makes little difference to the amount of reflected energy. What really counts is its radar reflectivity value, or cross-section, known as RCS. This is influenced by two factors: its shape, and the electro-magnetic properties of its materials. To digress for a moment, it has sometimes been stated that the Royal Air Force's Mosquito, the "wooden wonder" of the World War II, was fairly radar-resistant, due to its construction of birch plywood and plastic. The fact is that its huge whirling propellers made wonderful radar reflectors!

In practice, very large reductions have to be made in the RCS to have a significant effect on detection range; a simple order of magnitude will reduce detection range by 44 percent: worth having, but given the range of modern missiles not tactically very significant. Three orders of magnitude will give the really worthwhile reduction of 82 percent. But, as can be imagined, this is not easy to achieve.

Even as the shape of an airplane varies depending on the angle from which one looks at it, so its RCS also varies according to its aspect from the radar transmitter. Radars, both ground and airborne, will generally be looking outwards, while in most cases a threat aircraft will be inbound, although only rarely will they be directly aligned. Therefore the front quarter RCS is the most important area in which to make reductions. If the front of the aircraft can be shaped to deflect most of the radar energy off at an angle, significant

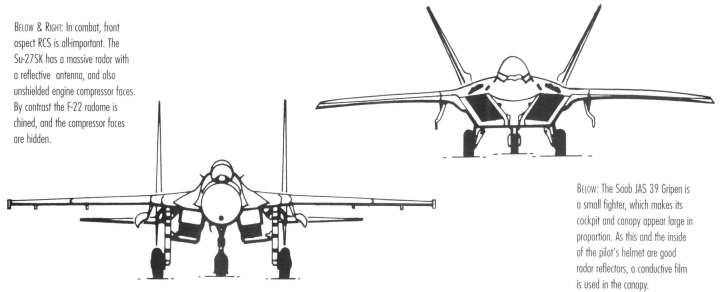

BELOW & RIGHT: In combat, front aspect RCS is all-important. The Su-27SK has a massive radar with a reflective antenna, and also unshielded engine compressor faces. By contrast the F-22 radome is chined, and the compressor faces are hidden.

BELOW: The Saab JAS 39 Gripen is a small fighter, which makes its cockpit and canopy appear large in proportion. As this and the inside of the pilot's helmet are good radar reflectors, a conductive film is used in the canopy.

savings to the RCS can be made. Under ideal conditions - i.e., when the aircraft is pointing directly at the radar emitter - the return will be good, but directly the angle changes, however marginally, it will be much weaker.

The trick involved in stealth shaping is to minimize the number of angles which give a strong radar return. This done, beyond a certain range, only occasional radar "paints" will be achieved as the angle between the emitter and the target changes, and these should be insufficient to allow any form of tracking or course prediction.

Stealth is extremely costly and difficult to achieve. Like all else it demands trade-offs, which may adversely affect performance, maneuverability, or weapons capacity. In many cases it may well be felt preferable to design-in only as much stealth as is needed to give a tactical edge over the projected threat, while maximizing conventional fighter capabilities.

Taking the frontal aspect first, the radar signature of external stores is excessive. For the greatest RCS reduction, all fuel and weaponry must be carried internally. This immediately makes for a large fighter. Another factor is the fighter's on-board radar, the antenna of which makes an excellent radar reflector. A fighter's radar is also an emitter, which if active immediately negates a covert approach. The ramifications of this will be dealt with in a later section.

Also up front are the engine intakes. The compressor face of a jet engine acts like a flat disc. As such it is an almost perfect radar reflector, and if a radar is allowed to look right down its throat it not only gets a perfect return, but can even identify the target type from the compressor harmonics.

To avoid this, the engine(s) must be located centrally, and fed by serpentine ducts, preferably located dorsally, even at the expense of thrust losses. Baffles and deflector vanes help

RIGHT: Locating the engine inlets above the wings and screening the serpentine ducts with carefully sized fine mesh eliminated one of the greatest contributors to RCS on the F-117A. The cockpit canopy was also made stealthy at the expense of pilot view.

BELOW: Lockheed used faceting to achieve stealth, by deflecting radar emissions off at an angle, but to the detriment of agility and performance. Consequently, the F-117A can operate only by night; it would be an easy target in daylight.

to contain the emissions, as does carefully sized fine mesh over the intakes. The intakes themselves must be simple fixed geometry; compression ramps and bypass doors are inherently non-stealthy.

The other big radar reflector from the frontal aspect is the cockpit, or even the inside of the pilot's helmet, seen by radar through his head! The answer here is a conductive layer on the windshield and canopy to carry the energy away. Have you ever noticed that F-16 canopies in particular have a slight golden tinge?

At this point it is time to consider radar absorbing material (RAM). This exists in two main forms: as tiles or as paint. The former uses rubber-based tiles or linings in cool areas, or ceramic tiles where heat is intense. Both these and paint are impregnated with iron oxides which are able to absorb microwave energy, converting it into minuscule amounts of heat. Widely used in and around engine intakes, and on the leading edges of wings, RAM is not a panacea, but it does help reduce radar returns, especially if used when all else has failed.

Small things, such as an ill-fitting panel, can add greatly to the RCS, and an extremely high standard of finish is required to avoid this. Wing (and tail surface) leading edges are good radar reflectors. RAM apart, not too much can be done about this, except to make a virtue of a necessity and align as many edges as possible with it - inlet lips, access

LEFT: Inbound and from head-on, the F-15C Eagle is deadly in close combat, but the instant it turns, its huge planform makes it visible from many miles away, thus ceding the advantage to smaller opponents still outside visual distance.

BELOW: "Top quality" stealth is largely unaffordable, while often performance is compromised. The Typhoon, seen here, is a trade-off between performance, agility, and low observability. The "smiling" chin intake is an unusual feature.

doors, etc. As we saw with the front section, this would concentrate radar returns accurately in a single direction. Of course, the same principle applies to trailing edges.

Beam RCS is the largest of all, but not only is this unavoidable, this aspect is the least vulnerable. Vertical surfaces must be avoided and, equally important, so must acute or right angles between plane surfaces. Rear aspect is inherently less stealthy than the front, but due to the engine effluxes this can hardly be helped. It must be remembered, however, that a very high speed retreating target is crimping in the effective range of a missile quite remarkably, reducing the importance of a stealthy rear aspect.

A degree of stealth shaping and RAM was used on the Lockheed A-12/SR-71 reconnaissance aircraft series, but the technology made its first major stride towards maturity with the F-117A Nighthawk. This used faceting to achieve radar stealth, at a measurable cost in performance. But as Ben Rich, former head of the Lockheed's famed Skunk Works, declared: "Computer capacity has reached the stage where we could even make the Statue of Liberty fly!"

Since then, times have moved on. The recently acquired ability to predict the RCS of compound curves led to the Northrop B-2A stealth bomber. But like the F-117A, the B-2A is strictly subsonic. In the foreseeable future, fighter design will remain a compromise between stealth and performance.

Above: Canadian F/A-18 Hornets routinely carry dummy cockpits painted on their undersides as an aspect deception measure. But the opinion of many is that if you are close enough to see it, you are too close to be fooled!

HEAT SIGNATURES

As infra-red lacks ranging capability, it is best used in conjunction with radar. It is widely used for missile target acquisition and homing, therefore low observability in this regime is an entirely sensible precaution. Modern IR-seekers "see" in two wavebands: 8-14 microns (a micron is one thousandth of a millimeter) where the atmosphere is at its most transparent to IR radiation, and 3-5 microns, where the radiation from engine efflux plumes is maximized.

The exhaust plume is an obvious target for an IR seeker, as is a hot engine nozzle. Mixing ambient air with the exhaust helps cool it rapidly, lowering the heat signature. Flattened nozzles as used by the F-117A assist this process by increasing the surface to volume ratio, therefore making mixing with the ambient air easier. Another obvious step is to locate the nozzles dorsally, where they are shielded from below. But if thrust-vectoring is required, this solution is a non-starter.

The use of supersonic speeds without afterburning, known as supercruise, also helps. Careful design can minimize hot spots on the airframe caused by kinetic heating, but not to any great degree.

VISUAL SIGNATURE

Size has a great influence on visual detection. Several years ago, at a USAF Aggressor Squadron gathering, the writer posed the question: what did they want to fly on their next assignment? The unanimous verdict was the F-16. Sure, the F-15 Eagle was a great fighter, but who wanted to be the largest target in the sky? The F-15 was fine inbound and head-on, but the instant it turned and showed its planform it became visible at about 8nm (15km) to pilots in smaller fighters which were still outside its visual range!

A tactical trick called "knife-edging" was at one time, and probably still is, widely taught. The speed of modern fighters means that when sparring for position they are often at the very limits of visibility. When this occurs, a rapid change of aspect from planform to elevation often causes the opponent to lose visual contact, and with it, the fight!

For many years, the use of air superiority gray paint was mandatory in lowering visual signatures, while the intakes of the F-4 Phantom were painted white in an attempt to reduce the "black nostril" appearance from head-on. Many other camouflage schemes were tried, although these were mainly in the field of deception rather than stealth. Canadian F-18 Hornets carried a dummy cockpit painted on their undersides to give aspect deception, although the general opinion was that, if you were close enough to see them, you were too close to be fooled. American aviation artist Keith Ferris evolved a scheme of countershading. This was effective when flying wings level at high noon, but did not work under other conditions. The argument for color schemes which give instant recognition in the heat of the dogfight (shades of Richthofen) retains a certain validity.

DETECTION SYSTEMS

The evolution of the fighter was driven by two main factors: the nature of the air threat, and the development of the weapons needed to counter it. Increased speeds and altitudes demanded ever longer killing ranges, which in turn put a premium on early detection and identification - or, to use the modern expression, to expand the pilot's situational awareness (SA) bubble. The latter can be visualized as a variably shaped balloon extending around the fighter. In all, four detection systems have been widely used: visual, radar, electro-optical, and infra-red.

VISUAL

Visual detection is very limited. The wavelengths of visible light are extremely short, which gives an extremely accurate and detailed picture of the target. This is excellent for giving instant information on precisely what an opponent is doing at any particular moment. But the weak link is the human brain, which makes it much less certain on identification, as witness the shooting down during the Gulf War of two friendly Blackhawk helicopters over Northern Iraq by two USAF F-15s. Nor is it very accurate at assessing distance. Its main limitation is short range; even under ideal conditions it is difficult to spot a small aircraft such as the F-16 in the head-on aspect at more than about 4nm (7.4km), although a few individuals with exceptional eyesight can do rather better than this. In less than ideal conditions, this figure halves. Nor can the human eye penetrate cloud or darkness, while even in daylight there are large blind spots below and astern of the aircraft, and beneath the wings. Visual acquisition is a fall-back stance, adopted when the fight closes to knife range.

Radar is the primary fighter detection system. Not only does it have all-weather capability, but it extends the SA bubble out to between 50-100nm (93-186km) in front of the fighter, albeit in a rather shallow pie-shaped section, typically 60 degrees on each side of boresight and 8 degrees in elevation. The latter can of course be increased by making the scanner "nod" up and down, usually 60 degrees up or 40 degrees down, but not both at once.

As mentioned earlier, radar gives very accurate range data, and is equally good at assessing target speeds, particularly when the rate of closure is high, as is the case with a head-on attack. However, it is less good at angular discrimination. This is due to the wavelengths used, which are considerably longer than that of visible light. Seen through the eyes of radar, the target would be a fuzzy blob, constantly growing and diminishing and changing shape, and with its epicenter sometimes moving off the target altogether in what is called glint.

This has two main effects. A close formation at any distance tends to show up as a single contact on the radar screen; this happened on several occasions during the Gulf War.

BELOW: The Boeing E-3 Sentry is most capable AWACS in the world, able to track multiple targets at long ranges and over vast areas. It allows friendly fighters to commence attacks while leaving their radars on "standby."

BELOW: The Ericsson PS-05/A radar gives multi-target tracking and raid assessment while continuing to scan, and mid-course guidance for Amraam or Mica. At center, air combat mode gives auto-acquisition and high resolution single target tracking.

BOTTOM: Pulse-Doppler radar works against low-flying targets by screening out all ground returns moving in conformity with the flight path. But if the target, warned by his RWS, turns to 90deg from the fighter, radar contact is all too easily lost.

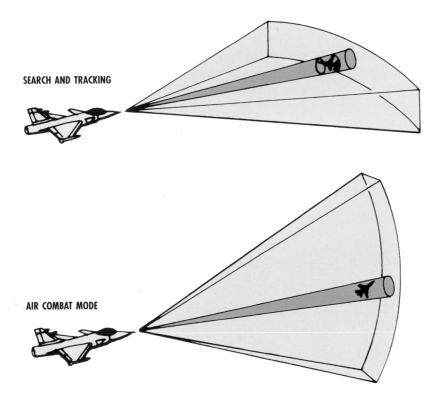

SEARCH AND TRACKING

AIR COMBAT MODE

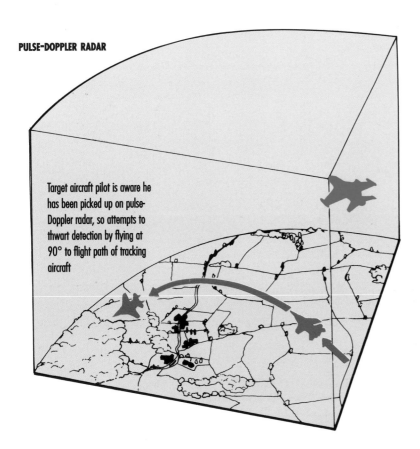

PULSE-DOPPLER RADAR

Target aircraft pilot is aware he has been picked up on pulse-Doppler radar, so attempts to thwart detection by flying at 90° to flight path of tracking aircraft

Various techniques have been developed to overcome this, but the problem persists. The other effect is that radar-homing missiles are inherently less accurate than IR types, which means that the lethal radius of the warhead must be large to accommodate a greater miss distance.

Between 1939 and about 1970, the pilot was forced to interpret raw radar material on his screen, and if the target was below him he had the almost impossible task of picking it out of the ground clutter. Things have since changed. Radar information is now presented in an easily understandable alpha-numeric format, while pulse-Doppler (pD) in the lookdown mode allows all returns moving in conformity with the flight path of the fighter to be screened out, leaving only valid contacts.

Modern fighter radars are multi-mode; some modes are optimized for air combat, others for attacking ground targets; yet more for navigation. There are normally several air combat modes; the designations and to a degree the functions of these vary according to the manufacturer. For example, let us consider the Westinghouse APG-68 as carried by the F-16C.

At the detection stage, Range While Search (RWS) mode gives the longest detection range over the widest area. Once contact is made, the pilot can designate a target. As soon as conditions are favorable, the radar then automatically switches to Single Target Track (STT), which continues until the interception is complete, or the attack is broken off.

An alternative detection mode is Track While Scan (TWS). This has a much smaller scan angle, but using a facility called Multi-Target Track (MTT) it creates files for each hostile contact, giving priority to the closest ones. As the range closes the F-16 pilot will probably go to Air Combat Mode (ACM), which covers both missile launch and gun firing parameters, with STT as an alternative.

Radar tracking is far from foolproof. It is possible, particularly when looking down, to lose a target without at first knowing it. A target turning through 90 degrees to the boresight of the fighter will for a short while be moving in conformity with the ground beneath the fighter, and the pD filters will screen it out. Alternatively, the target might make a violent maneuver which takes it outside the area of the scan. The problem arises because TWS mode is programmed to extrapolate existing tracks for up to 13 seconds, and will continue to display them on screen. The result is that the pilot will not immediately know that contact has been lost, and in air combat 13 seconds is a long time.

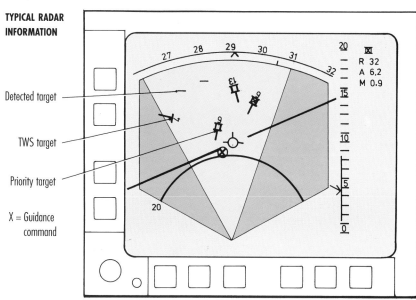

ABOVE: The combination of the F-14 Tomcat and the AIM-54 Phoenix caused consternation in the Warsaw Pact air forces when it entered service. The long-range multiple-kill capability had no precedent, and the Russians had no answer.

RIGHT: This is what a Gripen pilot sees on his radar screen: a detected target, a TWS target, a priority target complete with altitude and velocity vector, and a guidance command in the event of Amraam or Mica being launched.

TYPICAL RADAR INFORMATION

Detected target

TWS target

Priority target

X = Guidance command

For many years, radar allowed only one target to be attacked at a time. This limitation was far from satisfactory. The mold was broken by the F-14 Tomcat/AIM-54 Phoenix missle combination. The Phoenix, a long-range weapon, used active radar terminal homing, making the operational requirement of the fighter the ability to put a series of missiles within about 10nm (18km) of their respective targets, from where AR terminal homing could complete the interception.

Prior to launch, up to six targets were electronically tagged, and a Phoenix was allocated to each. The big missiles were than salvoed in quick succession. The Tomcat radar continued to track the multiple targets,

providing mid-course guidance updates during the missile's time of flight. The same principle has more recently been adopted for smaller medium range AR homers, such as the AIM-120 Amraam.

DESIGN CAPABILITIES OF F-14 TOMCAT/PHOENIX COMBINATION

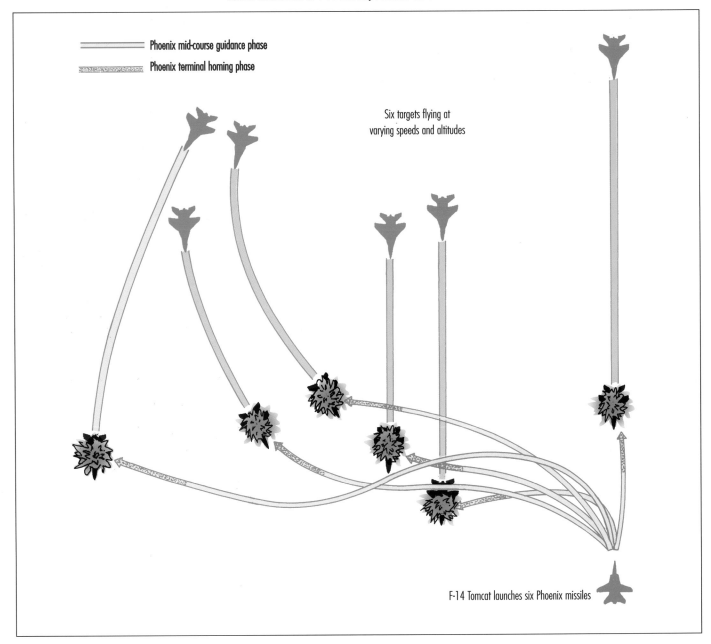

Phoenix mid-course guidance phase
Phoenix terminal homing phase

Six targets flying at
varying speeds and altitudes

F-14 Tomcat launches six Phoenix missiles

ABOVE: This diagram is indicative of the potential of the F-14/AIM-54 combination — six opponents down without being able to launch a single missile in return. But, as it happened, Phoenix was not used in anger until 1999, when it scored three misses against Iraqi planes.

Fighter radar has certain inherent disadvantages. Firstly, it is vulnerable to electronic countermeasures of various types. Secondly, it is an emitter, and as such betrays the position, heading and speed, and possibly intentions of the fighter, to the opposition. When it "locks on" to a target, which is generally known as attack mode, its signals change, and this is enough to warn an opponent of hostile intent. Finally, it only looks forward, and then through a very limited arc.

There are various remedies to cover against attack from astern. At the most basic, it is a wingman checking the vulnerable six o'clock area. The alternative is either ground radar or airborne early warning (AWACS) aircraft, which can monitor a vast area of sky around the fighter, and provide timely warning of any hostiles sneaking in behind.

ELECTRO-OPTICS

Electro-optical (EO) systems are more of an identification aid than a detection system, and reportedly can give a positive visual identification (VID) on a small fighter at about 20nm (37km). This is not dependent on perfect visual conditions; light enhancement techniques allow its use in semi-darkness, although of course it cannot see through cloud. It is therefore most effective against a long-distance hostile aircraft.

Another use was demonstrated over the Mediterranean on January 4 1989. A pair of US Navy Tomcats from the carrier *John F Kennedy* were harassed by two Libyan MiG-23s. Five times the Tomcats turned away, albeit at a shallow angle which allowed them to continue to track the MiGs on radar, but on each occasion the Libyans jinked back onto a collision course for interception. As the range neared the maximum for missile launch (by the MiG-23s), EO established that the MiGs were armed with Apex and Aphid missiles. A reaction could no longer be delayed; the Tomcats launched missiles, and both Libyans went down.

INFRA-RED

Infra-red sensors are used for search and track (IRST). The fact that they are non-emitting passive sensors means that the target does not know that it is under surveillance, but their use is limited by their almost total lack of ranging capability, although for short distances this deficiency can be made good by using them in conjunction with a laser ranger.

In the field of detection, IR sensors are more a supplement to radar. For example, when the radar is trained to look downward IRST can be used to scan the heavens above, to provide early warning that something is approaching. Because IR is only just outside the spectrum of visible light, its wavelengths are very short, which gives it much better angular discrimination than radar. Not only does this give greater accuracy in both azimuth and elevation, it can also pick out individual aircraft in close formation at ranges where radar shows only a single contact. However, it does have severe limitations in conditions of adverse weather, and cannot see through cloud.

DEFENSIVE SYSTEMS

The term "defensive systems" covers a whole gamut of sensors which detect threats of all types, direct and indirect, and the means of countering them. Let us first consider radar.

Of itself, radar is only an indirect threat, in that it does not actually shoot down aircraft. Having said that, it is used for detection and tracking, by both ground and airborne stations (AWACS), to aid interception, and it is used for missile homing (SAMs and AAMs). In an ideal world, they could all be jammed, but the sheer number of them, the spread of frequencies, and in many cases the power of the transmissions, make this impracticable. The compromise answer is to detect, analyse, and identify them, to at least give the fighter pilot some warning.

This takes the form of a radar warning receiver (RWR), otherwise known as an electronic surveillance measures (ESM) system. As we have seen, the power of electronic emissions decays rapidly with distance, and often, but not always, the RWR will detect the incoming signal rather earlier than the enemy receiver will be able to pick up its echo. By

measuring its frequency and several other parameters, and comparing them with a pre-stored library of known emitter types, the RWR can deduce its identity and flash up an on-screen warning, giving direction and distance, the latter being based on signal strength. The warning is usually accompanied by an audible tone, to attract the pilot's attention.

When the threat is direct – for instance, if a SAM is on its way, or an enemy fighter radar has just gone into attack mode, indicating that an AAM launch is imminent - countermeasures are called for. This may take one of two forms: chaff, or active jamming.

ABOVE: Electronic warfare is now so important that training has now become a fine art. Seen here on a Gripen is the Ericsson A100 Erijammer responsive system, designed to upgrade the skills of radar operators.

BELOW: Although the most dramatic thing about this picture is the moment of motor ignition of an AIM-9 Sidewinder, the real interest lies in the ALQ-131 ECM pod carried ventrally to give this F-16C 360deg all-round protection.

INVERSE GAIN DECEPTION

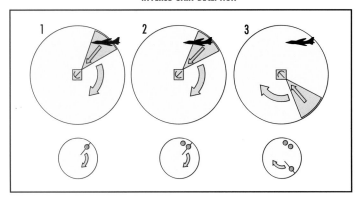

RANGE DECEPTION JAMMING

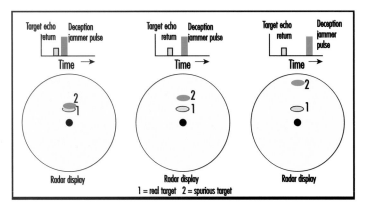

Chaff

Of these, chaff is the simplest, and most modern fighters carry chaff dispensers. Chaff consists of either metal foil or aluminum-coated glass, cut to a length of half the wavelength of the threat radar; the cutting is done at the time of release to match the radar to be defeated. Deployed in large clouds, it gives a false echo on enemy radar screens.

Chaff is not a panacea. Naturally it gives no protection against a head-on fighter attack, and it is of little use against continuous-wave radars. Once released, the chaff clouds soon fall behind the fighter, and enemy fighter radars in pulse-Doppler mode can quickly screen it out using the velocity differential.

Noise jamming

Active ECM takes two forms: noise jamming and deception jamming. The former is used to "drown out" the weak return from the target by transmitting a continuous signal over the frequency range of the enemy radar, but a lot of power is needed. A non-directional jammer, dissipating its energy through 360 degrees is unlikely to be effective. What is needed is a directional jammer, able to aim its emissions right down the throat of the hostile radar receiver for maximum effect.

ABOVE: Range deception jamming consists of amplifying and returning radar pulses while gradually increasing the delay to create a spurious target return. When the real target reaches the edge of the screen, both returns vanish.

RIGHT: The excrescence near the top of Rafale's fin houses SPECTRA ECM equipment, with forward and aft transmitting antennae. Just visible are the wingtip missile rails which house fore and aft RWR antennae.

This is fine if the radar to be jammed operates on a single frequency. If it is frequency-agile, however, directional jamming will not work. The answer then is barrage jamming, but this needs so much extra power and volume for black boxes that it can only realistically be done by specialized ECM aircraft.

DECEPTION JAMMING

Deception jamming is an ongoing battle between radar designers and ECM specialists, the advantage lying first with one, then the other. In essence, it consists of processing and amplifying the threat radar emission, then transmitting it in such a way that it is mistaken for the genuine article. Typical forms of deception jamming are inverse gain, and range-gate stealing, although there are many others.

Inverse gain deception jamming takes advantage of the fact that not all the energy of a radar emission goes in the direction intended; some spills over the edge in the form of sidelobes. By transmitting fake returns back to the receiver, of sufficient power to enter via the sidelobes, the radar may well accept these as genuine and reject the real echoes. By a careful sequence of timing, a multitude of spurious target returns can be created on completely different bearings.

Range-gate stealing is used against fire control radars. When the latter lock-on, what is known as a range-gate is used to concentrate on the target. The jammer receives the signal, amplifies it and, after a brief delay, re-transmits it. This has the effect of giving two contacts, with the second, more powerful one apparently at a greater distance. Consequently, the range-gate shifts to accommodate this. As the process continues, the apparent target moves steadily further away. Then when the jamming stops, the false target disappears, and the real target must be reacquired. Often this is no longer possible.

Jammers for fighters should be fully automated, controlled by software. While these are reprogrammable, the technology is now so complex that the widely used expression "it's only a software change" is now anathema. The jammers are ideally located within the airframe, with the antennae spaced to give all-round coverage. Pods are the alternative, but these sterilize a pylon, and compromise stealth, while all-round coverage suffers from airframe masking.

IRCM

Infra-red countermeasures are almost entirely designed to thwart heat-homing missiles. For the most part they take the form of flares, in the form of superheated ceramic "bricks"

which, launched in dozens, give the missile seeker head a plethora of alternative, equally attractive (and spurious) targets. But the really clever bit is knowing that an IR missile is on its way!

COCKPIT AND COMMUNICATIONS

The miraculous escape of Vyacheslav Averianov and his navigator from their stricken Su-30 at Le Bourget in June 1999 highlighted how far ejection seats have come since their operational debut in January 1943. As aircraft became faster, the difficulties of escaping from them multiplied.

Firstly, the canopy had to be jettisoned without harming the occupants. Then the seat had to be shot clear of the aircraft, fast enough to avoid the tail, but not so fast as to permanently damage the man. In an ejection, arms and legs tend to flail about; these had to be restrained. The Lightning was an extreme example; it was very short of clearance under the instrument panel, and thigh length was a critical parameter for its pilots.

BELOW: Martin Baker have led the world in ejection seat design for decades. Seen here is the Mk 10A seat as used in the Tornado F.3. The straps hanging down at the front are leg restraints.

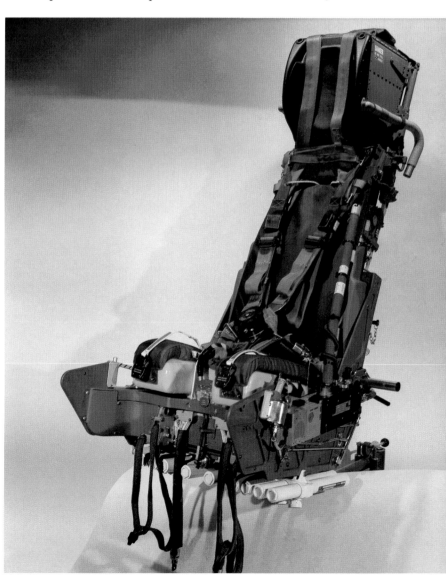

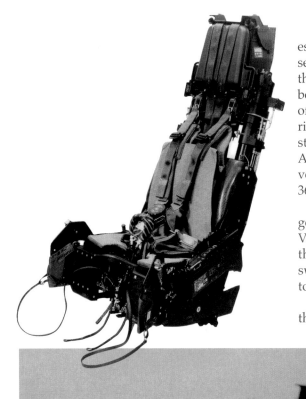

RIGHT: The Eurofighter Typhoon will use the Martin Baker Mk 16A seat which, as can be seen, differs appreciably from the Mk 10A. It is much lighter, less complex, and much more capable than previous ejection seats.

The next step was to provide a means of escape at low level; gradually the zero/zero seat was developed to allow ejection while the aircraft was stationary on the runway. But being able to eject at low level was not the end of the story. What if the aircraft was not the right way up at the critical moment? The next step was the vertical seeking seat. When Averianov ejected, the Su-30 was past the vertical and still fairly low. Yet his Zvezda K-36D seat saved him uninjured!

By the late 1960s, pilot workload was getting past a joke. In the skies over North Vietnam, many US pilots missed kills because they could not perform the necessary switchology to go from one weaponry setting to another in the time available.

A pilot normally flies with his left hand on the throttle and his right hand on the control

RIGHT: Cordite cartridges lift the Mk 16A seat into the air on test. The piece sticking out of the back gives aerodynamic stability, while the arms each side of the seat near the pilot's head are sensors, which tell the computer which way is up.

LEFT: First introduced on the Boeing F-15 Eagle, HOTAS (Hands On Throttle And Stick), was a means of putting all controls needed in critical phases of flight under hand, without the need to grope around in the cockpit to find the correct switch.

column. During the Vietnam War the need to let go one or the other while he selected flaps or speed brakes, switched radar modes, or whatever - or even worse, had to change hands to perform some actions - was not conducive to combat efficiency.

HOTAS

It was to overcome this that the HOTAS (Hands On Throttle And Stick) concept was pioneered in the F-15 Eagle. The HOTAS solution was to put everything required in combat, or in critical flight stages such as landing, on the throttle and stick. With nine switch functions located on the dual throttles and another six on the stick, operating HOTAS called for the manual dexterity of a piccolo player, and initially required a great deal of practice to achieve operational efficiency. On the other hand, it paid off. No longer did the pilot have to grope around the cockpit for the correct switch or lever while not daring to take his eyes off a distant target in case he lost sight of it. HOTAS has since become standard throughout the fighter world.

Almost equally revolutionary was the cockpit of the F-16, with its single piece cockpit transparency, its steeply raked seat, adopted to increase g-tolerance, and its sidestick controller. While the view "out of the

window" is unparalleled, the one piece transparency has not been copied. The steeply raked seat has been adopted elsewhere, although it has been criticised on two counts. One is that turning to look behind under high g-loadings is a potential cause of neck and shoulder strains. The other is that the raised knee-line reduces dashboard space for instruments. Nor has the sidestick controller, which again reduces console space, really caught on.

BELOW: Although very cramped, the cockpit of this MiG-21 has been modified with HOTAS, with multi-function displays, and with a wide angle HUD, by Israeli Aircraft Industries. This is the MiG-21 2000, seen at Le Bourget in 1995.

THE "GLASS COCKPIT"

The final trendsetter in cockpit technology was the F-18 Hornet. This also had a raked seat which reduced dashboard space. This lack was overcome by doing away with the mass of "steam-gauge" dials and instruments, and replacing (with the exception of a few back-up instruments) most of them with three color multi-function displays on the dash. These allowed the pilot to call up whatever information he wanted at the touch of a button and, combined with the Head-Up Display (HUD), which in various forms had been common to all fighters for many years, comprised the first "glass cockpit."

Over the past quarter century, cockpit design has been revolutionized by these advances. The advantages of HOTAS and the glass cockpit are so great that all fighter cockpits since have been designed on the same principle, while hundreds of older machines are to be updated in a similar manner.

COMMUNICATIONS

Communications between one fighter and another, and with ground control, have traditionally been made via VHF radio. But this is subject to enemy jamming; in any case, in the heat of battle there is little time for a pilot to absorb incoming information. The modern answer is secure data link, which can not only be used to carry speech, it can take tactical information straight through to the radar displays and superimpose it on the existing situation. This not only saves time, it gives the pilot an instant update at any given moment. As such, it is invaluable.

QUANTITY VERSUS QUALITY

In the 1970s, with worldwide inflation rampant, aviation writer Bill Gunston evolved what he called his economic disarmament theory. Extrapolating rising fighter costs, he arrived at a point where a single aircraft equaled the Gross National Product of even the most advanced nation. In the event of war, our fighter took off to engage the enemy fighter!

Fortunately things never did get that bad, but the central premise holds true: a state of the future art, gold-plated fighter is all but

ABOVE LEFT: Like the F-16, Rafale has a sidestick controller, liberally festooned with buttons and switches as are the throttle levers. One Dassault innovation, first seen on the Mirage 2000, is a look-level display directly beneath the HUD.

BELOW LEFT: Gripen has a "glass cockpit" with three color multi-function displays and a holographic Hud. The proliferation of buttons and switches on the stick and throttle indicate that Saab have adopted the HOTAS concept.

unaffordable. The US Navy was forced to soldier on with the F-14A Tomcat, powered by engines originally regarded as interim types pending a better one coming on stream. The US Air Force had developed the F-15 Eagle, at that time the ultimate fighter, but could not afford anywhere near as many as they wanted.

There is an old saying: numbers create confusion, and confusion degrades technology. Applied to fighter combat, it implies that even a force of superfighters can be overwhelmed and defeated by superior numbers of inferior aircraft, as happened to the German Me-262 force in 1944/45. There are historical precedents aplenty: remember Leonidas and his 300 Spartans at Thermopylae? Or, more recently, the Alamo? Custer's last stand?

The fact is that in war, unless the adversary is totally inferior as in the Gulf War of 1991, class is not enough. One needs numbers. Not numbers of aircraft on the ground, but numbers of aircraft in the air. The most generally anticipated threat during the 1960s was the Soviet Union, which deployed hordes of cheap and cheerful but mainly agile fighters which, in the event of war, would greatly have outnumbered NATO forces.

Having pondered the problem, the USAF came up with the hi-low mix: a nucleus of hi-tech and very capable fighters, backed by a larger number of austere, and therefore affordable machines, easy to maintain, and therefore possessed of high sortie rates.

This in turn led to the Lightweight Fighter Competition, which was won by the F-16. Developed in a "Skunk Works" type operation at Fort Worth, the F-16 was designed as an easily maintainable fighter, affordable in sufficient numbers to back up a nucleus of F-15s.

The historical record showed one interesting fact: the more heavily outnumbered the force, the better its exchange rate was in terms of victories per hundred sorties. But at the same time, its loss rate usually grew to unacceptable levels. There was obviously little advantage in a superior exchange rate if the war was lost! Also, having to fly what was widely considered to be an inferior fighter did nothing for pilot morale.

The forecast war between NATO and the Warsaw Pact in Western Europe never came to pass. The austere F-16 was upgraded into a very capable fighter by any standards, and the USAF ended with a rather better hi-low mix than it had anticipated.

ABOVE: Quantity versus quality. The Soviet Union and its allies operated hordes of MiG-21s and MiG-17s (nearest and third from camera). To reduce the numerical imbalance, the USAF produced the F-16 (second from camera).

THE FIGHTERS

Above: The Dassault Mirage 2000-5 blasts skywards carrying a full load of AAMs and drop tanks. The Mirage 2000 was optimized for the top right hand corner of the flight performance envelope, and is at its best at high altitude.

The term "fighter" has for many years been a much misused expression, applied to virtually every aircraft that carries a gun or other weapon that might conceivably be used against an aerial opponent. Even the Nighthawk, the stealthy stalker of the skies above Baghdad in 1991, carries an F for Fighter designation, even though the only offensive capability of the F-117A against other aircraft is for the pilot to make rude gestures from the cockpit. Always assuming that he can see them, since visibility "out of the window" is truly appalling.

The main criterion for the selection of fighters in this section is that they must have air combat - either air superiority or interception - as their primary function. The fact that they may have secondary roles as ground attack aircraft, and most have, is for our purposes basically irrelevant.

The second criterion revolves around the term "modern." An arbitrary cut line has been drawn at 1970, as it was this year that saw the genesis of fighters as weapons systems proper, rather than merely as weapons carriers. Of course, a handful of much earlier fighters, notably the F-4

Phantom, the MiG-21, and the Northrop F-5E, have been made viable in the present era by way of updates which include modern radars and avionics, HOTAS, a "glass cockpit," and more powerful and economical turbofan engines. Be this as it may, they are modernized rather than modern fighters, with outdated structures and aerodynamics, and as such have been excluded from the list that follows.

The aircraft data included hereafter with each individual type can give no more than vague clues to actual combat worth.

DIMENSIONS

Overall dimensions are given in feet and meters. These hint at the size of the visual signature only. Wing area, given in square feet and square meters, indicates the lifting area available, while aspect ratio, the ratio of span^2 to wing area, modifies both roll rate, and speed bleed-off during hard maneuvering. The lower the aspect ratio, the faster the roll rate but the higher the energy loss.

WEIGHTS

Weights are given in pounds and kilo-grammes (lb/kg). Empty weight is the brochure figure where this is available; otherwise it has been calculated. Take-off weight is that for the fighter configuration, with full internal but no external fuel, guns loaded, and the normal load of air-to-air weapons.

Above: Aware of the vulnerability of conventional airfields, Sweden frequently operates from widened stretches of road. Here Gripen minimizes its landing run using maximum deflection of its fore-planes for aero-dynamic braking.

BELOW: This F-15B Eagle two-seater is powered by two Pratt & Whitney F100 afterburning turbofans. Note that the "turkey feathers" to the nozzles are absent, a frequent modification on early Eagles. The two-seater is fully combat-capable.

PERFORMANCE NOMOGRAM
V_{max} speed range Hi/Lo (Climb rates and ceilings similar)

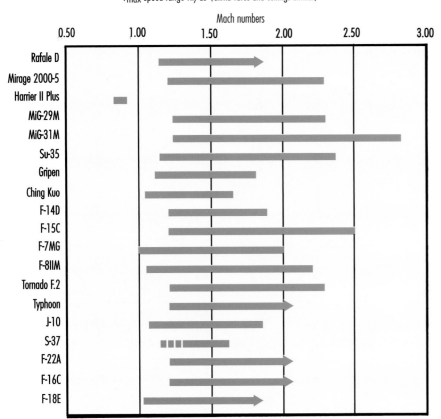

Mach numbers

| 0.50 | 1.00 | 1.50 | 2.00 | 2.50 | 3.00 |

Rafale D
Mirage 2000-5
Harrier II Plus
MiG-29M
MiG-31M
Su-35
Gripen
Ching Kuo
F-14D
F-15C
F-7MG
F-8IIM
Tornado F.2
Typhoon
J-10
S-37
F-22A
F-16C
F-18E

POWER
This includes engine(s) make and type, with maximum and military thrust stated separately in lb (kg). These outputs are of course for static thrust at sea level only.

FUEL
This is given in lb(kg) and is for internal fuel only. Where only volumes are available, it has been calculated as for JP-4 at 6.5lb/US gal. There is little point in including the weight of external fuel, as this could in any case be supplemented by in-flight refueling. Fuel fraction is expressed as the proportion of internal fuel weight at take-off to take-off weight. The optimum figure without compromising performance and agility is 0.30.

LEFT: This nomogram shows the difference between maximum speeds at high and low altitude. Arrows at high Mach numbers indicate that this figure can almost certainly be exceeded. Harrier II Plus is the odd one out; having no afterburner, it is not supersonic in level flight.

BELOW: Single and two-seat Typhoon prototypes DA 2 and DA 4. Compared to most canard deltas, the canards are set well forward to give the greatest possible moment arm. This has necessitated strakes under the cockpit to energize the airflow over the wings.

LOADINGS

Thrust loadings are expressed as a ratio of static thrust against weight, and are an indicator of specific excess power during hard maneuvering, and also acceleration and initial rate of climb. Thrust loadings exceeding 1:1 are today almost obligatory. Wing loadings, given in lb/sq.ft (kg/m²) are an indicator of instantaneous turning ability; combined with thrust loading, of sustained turning ability.

PERFORMANCE

Maximum speed, or V_{max}, is given by Mach number, both at high altitude and at sea level. Mach 1 is actually 88kt (163kmh) faster at altitude. Service or operational ceiling is given in ft/m, while initial climb rate, generally attainable at Mach 0.9 at sea level, is stated in ft/min (m/sec). Climb rate is also an approximate indicator of acceleration. All other data – rate and radius of turn, suitably qualified, and radar capability – are given in the text where known. Three things must be remembered. The performance data given are brochure figures, and for security reasons may well be inaccurate. In combat, no fighter even approaches its performance maxima. Finally, the ultimate arbiter in air combat is the pilot, not the aircraft.

WEAPONRY

This specifies gun types with caliber, and rate of fire and ammunition capacity where known, and maximum AAM load with types and homing systems.

ABOVE: Asraam, the next European dogfight missile, is test-fired by an F-16 from Eglin AFB. The small pictures record the destruction of a QF-4 Phantom drone. Asraam has proved very successful in trials to date.

LEFT: Hornets of Marine Squadron VMFA-312, clearly showing the long wing leading edge extensions. Also visible towards the rear of these are small metal angles, planted on to break up airflow effects which were overstraining the fins.

People's Republic of China
CHENGDU F-7MG

Developed from the Russian MiG-21F-13, at first sight the Chengdu F-7 would hardly seem to qualify as a modern fighter. But this would be to overlook the fact that the series has undergone a continual process of redesign, to the point where it bears only a passing resemblance to the original aircraft. Not only is the latest F-7MG model still in production, other variants, which may well enter service in the 21st Century, are still emerging in prototype form.

The F-7MG retains the pitot-type nose intake with translating shock-cone of the MiG-21. Like the Russian fighter, the radar antenna is housed in the shock-cone, the limited volume of which constrains the size of the radar. This is the multi-mode GEC Marconi Super Skyranger, which gives a true, if limited-range, all-weather capability.

The main distinguishing feature of the F-7MG is a cranked delta wing with increased span and area. The trailing edge has a slight forward-sweep outboard, and automatic maneuvering flaps. Stressed for +8g, it can sustain a turn rate of 14.7deg/sec at Mach 0.7 at sea level, but this falls away to 9.5deg/sec at Mach 0.8 and 16,405ft (5,000m).

The quest for greater capability led to the Super-7. With the aid of the American Grumman company, the F-7 Airguard was redesigned to have a solid nose to house the Westinghouse APG-66 multi-mode radar. The traditional Russian poor visibility from the cockpit was improved with a one-piece wrap-around windshield and a single-piece canopy.

A Western afterburning turbofan was proposed, with General Electric's F404 as the front-runner, although the RB.199 was also considered. It was fed by simple cheek intakes which precluded a Mach 2 capability, but allowed a handy Vmax of Mach 1.8.

Internal fuel capacity was increased; up to four AAMs could be carried; and wing loading was held down by a larger span and area. But in the aftermath of the 1989 Tiananmen Square massacre, US participation was terminated, and the project ground to a halt.

With Russian (MiG-MAPO) assistance, the Super-7 was revived and redesignated the FC-1, which was powered by the Klimov RD-93 afterburning turbofan rated at 17,985lb (8,158kg) maximum. The main external differences were sharply swept strakes to the wing leading edges which extended into shelves along the rear fuselage which carry control runs and the horizontal tail surfaces. The vertical tail also appeared to have undergone a radical change in shape.

As of early 1999, the radar for the FC-1 had not been selected, but the multi-mode pulse-Doppler GEC Marconi Blue Hawk, with a maximum detection range of 44nm (82km), is well in the running. First flight of the FC-1 was reported to have taken place early in 1997, with service entry expected in 2002, but these dates are now suspect.

This should have been the end of the line for the F-7, but in June 1998 a new technology demonstrator made its first flight. The F-7FS features a chin intake with a very obvious splitter plate, surmounted by an ogival radar nose, and is scheduled for 22 months of flight testing. Whether problems have been encountered with cheek intakes on the FC-1, or whether this is a "belt and braces" back-up solution, is not known. Only the Chinese know, and they are hardly likely to tell.

ABOVE: The Super-7, seen here in model form was the first radical attempt to upgrade the original F-7, but it never flew. It was followed by the broadly similar FC-1 which, while it is reported to have flown in 1997, is not expected to enter service.

LEFT: The F-7MG, seen here, has reverted to the original pitot nose intake. It differs mainly in having a double delta wing planform and revised horizontal tail surfaces.

F-7MG

DIMENSIONS: Span 27ft 2½in/8.32m; Length 48ft 10in/14.885m; Height 13ft 5½in/4.103m; Wing Area 268sq.ft/ 24.88m²;

ASPECT RATIO 2.76

WEIGHTS: Empty 11,667lb/5,292kg; Takeoff 16,623lb/7,540kg

POWER: 1xLiyang WP-13F turbojet; Maximum Thrust 14,550lb/6,600kg; Military Thrust 9,920lb/4,500kg

FUEL: Internal c4,500lb/2,041kg; Fraction c0.27

LOADINGS: Thrust (max) 0.88lb/lb-kg/kg; Wing 62lb/sq.ft/303kg/m².

PERFORMANCE: V_{max} high Mach 2; V_{max} low Mach 0.98; V_{min} 113kt/210kmh; Operational Ceiling; 57,417ft/ 17,500m; Rate of Climb; 38,388ft/min-195m/sec

WEAPONRY: 2x30mm cannon with 126 rounds 4xAIM-9P, R550 or PL-7 IR homers

USERS: (all types). Albania, Bangladesh, China, Egypt, Iran, Iraq, Pakistan, Sri Lanka, Sudan, Tanzania, Zimbabwe.

People's Republic of China
SHENYANG F-8IIM

ABOVE: The F-8IIM Finback is the best Chinese attempt so far to produce a capable all-weather fighter with adequate endurance. Ground clearance for the ventral tank looks marginal.

The People's Republic of China entered the jet era on the back of Russian technology. It license-built and, when relations cooled, reverse-engineered, the MiG-15, MiG-17, MiG-19, and the MiG-21. But these were all short-legged, clear-air day fighters, with little or no night or adverse weather capability.

Something more capable was obviously needed, but Chinese resources and technology during the 1960s and 1970s did not allow a move towards a radically different design. Instead, they chose to scale up the J-7 (MiG-21) into a much larger twin-engined aircraft, retaining the pitot-type nose inlet and translating shock-cone, the latter being large enough to house a more substantial radar scanner.

The result was the F-8, which retained the tailed delta configuration of the F-7, but with a slightly sharper sweep to its delta wing. Externally it bore a marked resemblance to the rather earlier Mikoyan Ye-152 prototype interceptor, which gave rise to speculation that it was actually the same design. But whatever the similarities, the F-8 was dimensionally rather larger and also differed in detail, which makes this unlikely. (In the Soviet Union the Ye-152 was abandoned in favor of the Sukhoi Su-15.)

The development period of the F-8, which was given the NATO reporting name of Finback, was unusually protracted. Although the maiden flight of the first prototype took place on July 5 1969, the first pre-production aircraft did not follow it into the air until April 24 1981.

It soon became apparent that an even larger and more capable radar was needed. To accommodate this, the F-8I was given a massive nose upgrade to become the F-8II. According to *China Today: Aviation Industry*, 70 percent of the aircraft was redesigned, although, nose and inlets apart, the basic layout was retained.

In the original machine, the engines had been close together, to minimize profile drag and to suit the shared nose inlet. In the revised F-8II, conflicting needs caused design compromises. The radome needed to accommodate the largest available antenna, but was limited by cockpit width, which in turn was limited by the need to minimize curvature to the side intakes.

These compromises became apparent on the occasion of Finback's only appearance in the West, at the Paris Air Show at Le Bourget in 1989. The side inlets, with large splitter plates and automatically variable wedges, were flared dramatically outwards around the flat-sided cockpit, which by US standards was cramped. From this point the forebody gradually widened into the solid radar nose.

Cockpit instruments were of the old-fashioned "steam-gauge" variety, and the radar display, set high on the right of the fascia, was hooded to assist viewing in strong sunlight. Forward view through the optically flat windshield was restricted by heavy framing, and a heavy canopy bow did little to help matters. Rear visibility was poor.

The wing was of very thin section, with conical camber towards the tips. Small wing fences were located outboard. The most remarkable thing about the F-8II was the ratio of its length to its span, which approached that of the USAF's F-101 and F-104.

A move to fit the F-8II with Western radar and avionics was canceled, and the Russians moved in by default. The latest variant, the F-8IIM, made its maiden flight in 1996, equipped with the Phazotron Zhuk-8 multi-mode pulse-Doppler radar. Its cockpit is still largely old-fashioned, but has a couple of multi-function displays, and a move towards HOTAS.

The F-8IIM is not the most maneuverable fighter around, but it can sustain a turn of 6.9g at 3,200ft (1,000m) and Mach 0.9, while it can accelerate from Mach 0.6 to Mach 1.25 at 16,405ft (5,000m) in just 55 seconds. It is reported to have a rapid roll rate, but to be sluggish in pitch.

F-811M

DIMENSIONS: Span 30ft 8in/9.344m; Length 70ft 2in/21.39m; Height 17ft 9in/5.41m; Wing Area 454sq.ft/42.2m²

ASPECT RATIO 2.07

WEIGHTS: Empty 22,864lb/10,371kg Takeoff 33,704lb/15,288kg;

POWER: 2xLiyang WP13B turbojets; Maximum Thrust 15,432lb/7,000kg each; Military Thrust 10,597lb/ 4,807kg each

FUEL: Internal 9,259lb/4,200kg; Fraction 0.275;

LOADINGS: Thrust (max) 0.92lb/lb-kg/kg; Wing 74lb/sq.ft-362kg/m²;

PERFORMANCE: V_{max} high Mach 2.20 V_{max} low Mach 1.06 V_{min} 162kt/300kmh; Operational Ceiling 59,058ft/ 18,000m; Rate of Climb 44,097ft/sec-224m/s

WEAPONRY: 1x23mm twin-barrelled cannon with 200 rounds; 2xPL-2B IR homers 2 or 4 xPL-7A SARH homers

USER: People's Republic of China.

ABOVE: The tailed delta layout is seen to advantage as the F-8IIM takes off, but with the People's Republic license-building the Su-27, its future looks uncertain.

European Consortium
PANAVIA TORNADO F.3

ABOVE: Tornado F.3 was developed during the Cold War as a dedicated interceptor to patrol far out over the North Sea in search of Soviet bombers and reconnaissance aircraft.

During the Cold War, the air defense of Great Britain posed very specific and fairly unique problems. Initially the threat was of nuclear attack, and an uneasy peace was maintained by deterrence in the form, among other means, of the RAF V-bomber fleet. The defensive priority was to protect the V-bomber bases for long enough to allow them to get off the ground. This mission was carried out by the Lightning, a very high performance, if short-legged, point defense interceptor. After several years, the Lightning was reinforced by the American-built F-4 Phantom. While the latter could generally be outflown by the Lightning in close combat, it was a far superior weapons system, with a beyond visual range kill capability, a second crew member to handle the complex radar and avionics, far greater endurance, and eight AAMs to give four times the combat persistence of the British fighter. Whatever its failings, the Phantom was arguably the greatest fighter of its era.

Gradually the threat emphasis shifted from an all-out global nuclear exchange to a large-scale and more or less conventional war in Western Europe. In this scenario, Britain would play a vital role as a staging post for US reinforcements - what has been called an unsinkable aircraft carrier. The conventional, or even tactical, nuclear threat then became two-fold: Soviet supersonic bombers raiding the islands directly, and long range, missile-armed aircraft sneaking around far to the north to attack the vital supply convoys. These had to be intercepted as far out as possible.

Fighter combat was not a valid consideration; the agile Soviet air superiority fighters had not the range. What was needed was a bomber destroyer, a dedicated interceptor carrying a full bag of AAMs, able to range out hundreds of miles from its base in adverse weather conditions, and operate autonomously in the face of intense ECM. A very advanced radar and weapons system was needed, with a two-man crew to share the workload. The requirement was eventually formulated as being two hours on station at 400nm (740km) from base. Much later this was amended to three hours on station at 300nm (556km) from base. Also required was the ability to use afterburner for extended periods in order to make the interceptions more certain.

The first step was to assess various US fighters. The F-14 Tomcat was in many ways suitable, but its TF30 turbofans were unreliable and its avionics were at that time considered dated. The F-14 was also unaffordable.

TORNADO F.3

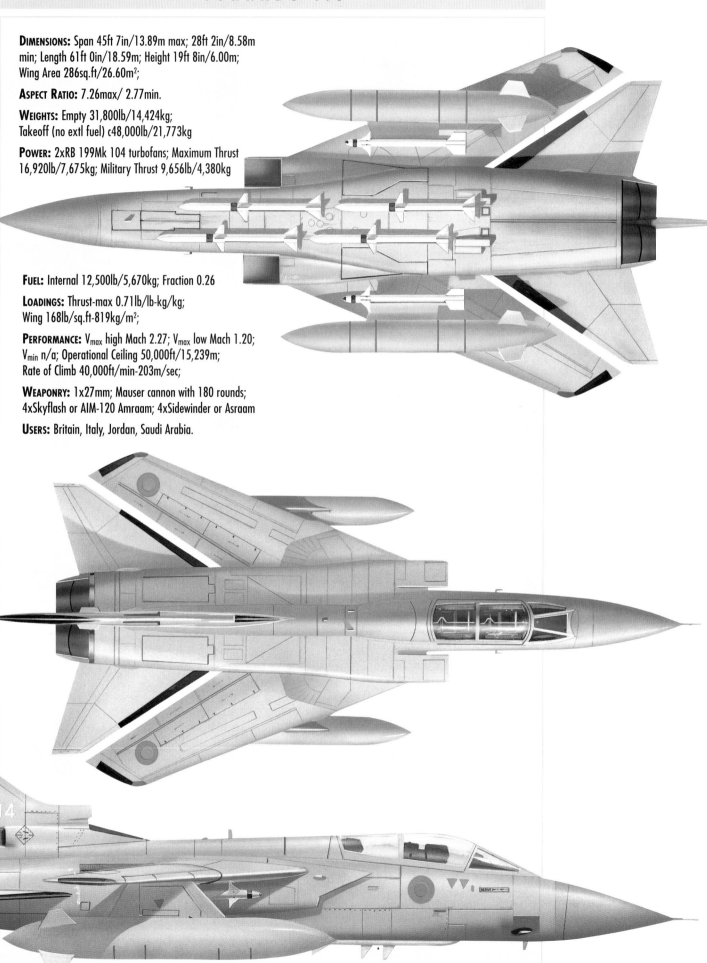

DIMENSIONS: Span 45ft 7in/13.89m max; 28ft 2in/8.58m min; Length 61ft 0in/18.59m; Height 19ft 8in/6.00m; Wing Area 286sq.ft/26.60m²;

ASPECT RATIO: 7.26max/ 2.77min.

WEIGHTS: Empty 31,800lb/14,424kg; Takeoff (no extl fuel) c48,000lb/21,773kg

POWER: 2xRB 199Mk 104 turbofans; Maximum Thrust 16,920lb/7,675kg; Military Thrust 9,656lb/4,380kg

FUEL: Internal 12,500lb/5,670kg; Fraction 0.26

LOADINGS: Thrust-max 0.71lb/lb-kg/kg; Wing 168lb/sq.ft-819kg/m²;

PERFORMANCE: V_{max} high Mach 2.27; V_{max} low Mach 1.20; V_{min} n/a; Operational Ceiling 50,000ft/15,239m; Rate of Climb 40,000ft/min-203m/sec;

WEAPONRY: 1x27mm; Mauser cannon with 180 rounds; 4xSkyflash or AIM-120 Amraam; 4xSidewinder or Asraam

USERS: Britain, Italy, Jordan, Saudi Arabia.

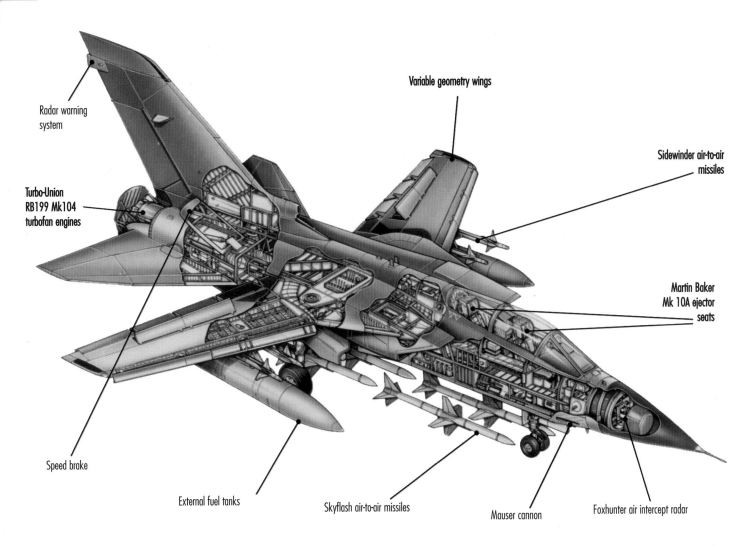

Radar warning
system

Turbo-Union
RB199 Mk104
turbofan engines

Speed brake

External fuel tanks

Skyflash air-to-air missiles

Mauser cannon

Foxhunter air intercept radar

Variable geometry wings

Sidewinder air-to-air
missiles

Martin Baker
Mk 10A ejector
seats

ABOVE: Changes from the original low level penetrator variant to suit it for long range patrol were minimal. The two-man crew, economical RB199 turbofans, and variable-sweep wing all proved assets for the new role.

The F-15 Eagle lacked range, and its avionics fit and single-man crew were assessed as being inadequate for the mission conditions. The solution adopted was to develop an interceptor from the existing tri-national Tornado interdictor/strike aircraft.

Although the baseline Tornado interdictor lacked many of the traditional fighter qualities, it had sound virtues of its own. Its RB199 augmented turbofans were designed for low specific fuel consumption at subsonic cruising speeds; while they lacked the poke needed by a fighter, and thrust dropped off alarmingly at speed in military power, they were well suited for an extended endurance mission.

The wings, so important for fighter maneuverability, featured variable-sweep, but this was to give a combination of economical cruising, good short field performance, reduced drag for a Mach 2 dash speed at altitude, and a smooth ride at low level. The first three were entirely compatible with extended endurance, while the last hardly mattered.

A major redesign was needed for the interceptor variant. The RB199 turbofans were given a 14in (36cm) extension aft of the flameholders to give extra burning volume. This not only gave a significant drag reduction: it gave added thrust when afterburner was used.

The main air-to-air weapon was the Skyflash missile, a British variant of the Sparrow with monopulse semi-active radar homing. Four were required, and to minimize drag they needed to be carried semi-recessed beneath the fuselage. As Tornado was quite a small airplane, this caused problems. The only way to do this was to lengthen the fuselage by inserting an extra bay behind the cockpit. Even then, the big missiles had to be staggered laterally to prevent their fins from overlapping. In addition, up to four Sidewinders could be carried on underwing pylons.

The nose was lengthened and reshaped to accommodate a new radar, the GEC-Marconi multi-mode pulse-Doppler AI.24. These alterations shifted the center of gravity and, to move the center of lift to compensate, the leading edge angle of the wing glove was changed from 60 degrees to 68 degrees. Another effect of the lengthening was to increase the fineness ratio, with a corresponding drag reduction which improved acceleration.

The additional bay was used to house extra avionics, and to increase internal fuel capacity. In fact fuel fraction on Tornado is a bit on the low side, and drop tanks, which more than double the internal capacity, are routinely carried. While these obviously degrade performance, in a mission where loiter time and range are of primary importance, this matters little. The "jugs" can be jettisoned at need, and as it is equipped with a retractable flight refueling probe, Tornado F.3 can always top up from a tanker in extremis.

The interceptor variant was given an improved flight control system, with automated fully variable wing sweep which the clever electrons scheduled according to speed and angle of attack. What could not be altered was wing area; this was relatively small, making wing loading very high for a fighter. Automatic flaps and slats helped to compensate by providing extra lift in the maneuvering regime, and a spin and incidence limiting system (SPILS), was fitted to provide carefree handling by preventing inadvertent departure. No longer did the pilot have to worry about straying over the limits.

The prototype Air Defence Variant, as it was originally known, first flew in October 1979, and entered service as the F.2 towards the end of 1985. Since then it has been upgraded to F.3 standard. Initially the radar failed to perform as advertised. It was designed to detect fighter-sized targets at low level at up to 100nm (185km). At closer ranges it was required to track up to 40 targets, while displaying the ten greatest threats, but

became unpredictable when doing this, with uncommanded switches into memory mode. When switched back, the tracks had often been lost. This is possibly largely the fault of the Cassegrain antenna. Selected as being more resistant to jamming, it appears to let too many false returns in via the sidelobes. Even in 1999, reliability remains a problem.

The Tornado F.3 navigator does rather more than monitor the avionics systems. In action he acts as a battle manager, advising the pilot of the best course of action. In this he is aided by data link, which feeds extra information through to his displays. Another use for Tornado F.3 is as a mini-AWACS. In conjunction with a gaggle of Hawks, the Tornado can throw the opposition into a state of confusion with a BVR attack, then direct the light fighters into an advantageous attacking position.

Tornado F.3 is not the most agile of fighters, but as it was designed that way it should not be criticized for this lack. In fact, when it was routinely used for air combat maneuvering, it quickly showed the strain, and had to be beefed up. It was used in a defensive role during the Gulf War of 1991, but operational circumstances prevented it from firing a shot in anger. It has since taken part in patrols over Bosnia, while the Italian Air Force has rented a couple of dozen pending the service entry of Eurofighter EF 2000.

BELOW: The fuselage of the F.3 had to be stretched in order to accommodate four Skyflash AAMs semi-recessed beneath it. Even then, they had to be staggered, as seen here.

European Consortium
EUROFIGHTER EF 2000 TYPHOON

ABOVE: The EF 2000 Typhoon is due to enter service from 2002, and is the eagerly awaited future fighter for four major European nations: Britain, Germany, Italy, and Spain. It has also aroused a great deal of export interest elsewhere.

Many years ago, a serving RAF pilot, writing in the magazine *Air Clues*, praised the interceptor Tornado for its single mission capability but lamented, "When are we going to get a *real* fighter aircraft?"

He had a point. A dedicated interceptor is all very well, but in the event of the "worst case" scenario of conventional war between Warpac and NATO in Western Europe, Tornado would almost certainly have been called upon to indulge in close combat against agile Russian fighters. This was not an inspiring thought.

The US had already recognized this; even the F-14 Tomcat, that lethal long range interceptor, had been designed for agility in the dogfight. The USAF were equipping with the superb F-15 Eagle, and the most agile close combat fighter of them all, the F-16 Fighting Falcon. The dual role F/A-18 Hornet was not

EF 2000 TYPHOON

DIMENSIONS: Span 35ft 10$\frac{3}{4}$in/10.95m; Length 52ft 4in/15.96m; Height 17ft 4in/5.28m; Wing Area 538sq.ft/50m^2;

ASPECT RATIO 2.40

WEIGHTS: Empty 21,500lb/9,752kg; Takeoff c33,730lb/15,300kg;

POWER 2xEurojet EJ 200 augmented turbofans; Maximum Thrust 20,250lb/ 9,185kg each; Military Thrust13,500lb/6,124kg each.

FUEL: Internal c9,921lb/4,500kg; Fraction 0.294

LOADINGS: Thrust (max) 1.20lb/lb-kg/kg; Wing 63lb/sq.ft-306kg/m^2.

PERFORMANCE: V_{max} high Mach 2 plus; V_{max} low Mach 1.20; V_{min} c130kt/241kmh; Operational Ceiling c60,000ft/18,287m; Rate of Climb c50,000ft/min-254m/sec ;

WEAPONRY: 1x27mm Mauser cannon with 150 rounds; 4xSkyflash or Amraam 4xSidewinder or Asraam

USERS: Britain, Germany, Italy, Spain.

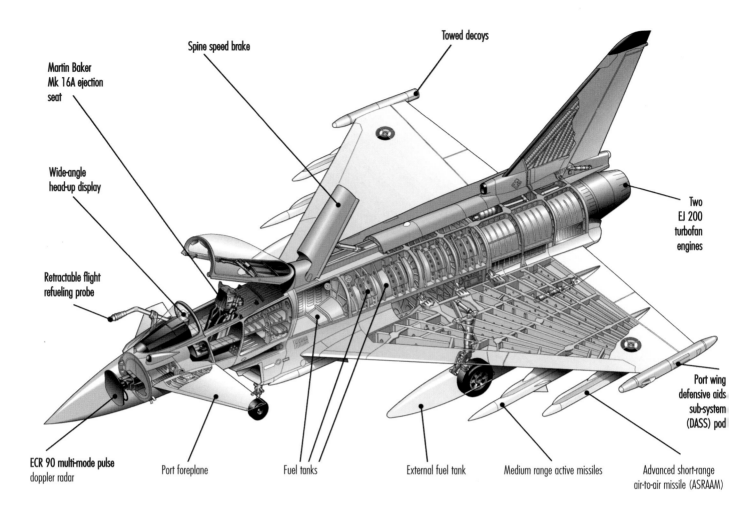

Martin Baker
Mk 16A ejection
seat

Spine speed brake

Towed decoys

Wide-angle
head-up display

Two
EJ 200
turbofan
engines

Retractable flight
refueling probe

ECR 90 multi-mode pulse
doppler radar

Port foreplane

Fuel tanks

External fuel tank

Medium range active missiles

Port wing
defensive aids
sub-system
(DASS) pod

Advanced short-range
air-to-air missile (ASRAAM)

LEFT: The EF 2000's "smiling" chin intake was adopted as a stealth measure. Also visible to the left of the cockpit is the IRST seeker which can be slaved to the radar.

ABOVE: Built round two powerful turbofan engines, the EF 2000 is extremely compact, with no wasted space. A very high proportion of the structure is made of advanced composites.

far behind for the US Navy and US Marine Corps. On the other side of the hill, two new Russian fighter prototypes had been observed at the flight research center at Zhukovsky in 1977. These later proved to be the extremely agile and formidable MiG-29 and Su-27.

In Europe, smaller nations were equipping with the F-16. But while this was affordable, it had no BVR combat capability, and in any case was expected to double as a bomb hauler. Of the others, the RAF deployed the Lightning and Phantom in Germany; the Luftwaffe used the Phantom in the air defense role; the main French fighter was the Mirage F.1, shortly to be replaced by the Mirage 2000; Italy's air defense fighter was the F-104S Starfighter, which in close combat turned like a gravel truck. Against the next generation of Soviet fighters, something better was clearly needed.

The obvious solution would have been for the smaller NATO nations to buy the F-15, but to do so would have meant abdicating advanced fighter design to the United States, thus giving the Americans a technological lead which could never be overtaken. Not only was this close to unaffordable in sufficient numbers, it was both nationally and politically unacceptable.

THE EUROPEAN SOLUTION

Fighter development is an extremely costly process, but the bigger the production run, the more units there are against which development costs can be amortized, which improves affordability. The obvious move was therefore for several nations with similar requirements to share in the development costs, offsetting these by a share in manufacture. But, when one considers that Britain's RAF and Fleet Air Arm were unable to agree a common specification for the Hawker P.1154 Mach 2 vertical takeoff fighter, getting four nations (rather than services) to agree on a common requirement is far more difficult. Even when the political commitment is present, it is a time-consuming process, as had earlier been demonstrated by the failure of the Anglo-French Variable Geometry Strike Aircraft, and later the Trinational Tornado interdictor/strike program. And time is money. Lots of it!

Feasibility studies did commence in 1983, and design weight was confirmed at about 10 tons. This pleased all but the French who favored an altogether lighter machine. Eventually they pulled out, to develop the *Avion de Combat Experimentale*, which later became Rafale. The remaining three nations, joined in 1985 by Spain, which needed to replace its Phantoms and Mirage F.1s, continued with the project. By December of that year, they had produced a European Staff Requirement, ESR-D, covering all requirements from combat to maintainability. The general layout was agreed; the new fighter was to be a twin-engined unstable canard delta; and the Eurofighter consortium was set up to develop and produce it.

Fighter development is a high risk process; to compound this, separate consortia were formed to develop dedicated engines (Eurojet EJ 200) and radar (European Collaborative ECR 90) for the new fighter. Both were equally high risk.

A twin-engined layout carries built-in headwinds: added volume, added complexity, and added structural weight to house them. Operational and performance requirements demanded a medium-weight fighter. As there was no engine in sight with a thrust rating adequate for the task, the penalties of a twin-engined layout were accepted. The good news was that the fuselage width meant that

four large BVR AAMs could be carried semi-conformally. A lesser consideration at the time was that two engines provided an extra margin of safety, but even then reliability was rapidly reaching the stage where any difference became marginal.

At that time there was no European turbofan in the right thrust class. The General Electric F404 was about the right size, but it was American, and it was nearly 20 years old, with little development potential. The basis of what became the EJ 200 was the experimental Rolls-Royce XG40.

Radar posed the same problem. To buy a radar from abroad was to abdicate the world radar scene; the choices were the USA or France. A new, up-to-the-minute European radar was badly needed, especially in the aftermath of the potentially great but flawed AI 24. Both politically and technically, the decision to go with the ECR 90 was the only possible one.

DESIGN CONSIDERATIONS

The requirements of modern combat are firstly BVR, in which the fighter detects early, accelerates to supersonic speed to give maximum energy to its missiles, then turns away hard to escape from the envelope of any enemy missiles which might have been launched in their turn, while deploying ECM to defeat them. This is essentially interceptor mode, for which Tornado F.3 was designed. However, the record was less than convincing. Taking into account the Indo/Pakistan conflict, the various Arab/Israeli wars, and Vietnam, the number of BVR victories could

ABOVE: From this angle, the long moment arm of the canard foreplanes is readily apparent. Also visible are the flaps and slats which provide variable camber.

be counted on the fingers of one hand! Nor, although this could not be known at the time, would there be many in the Gulf War of 1991.

As Korean War ace Frederick "Boots" Blesse commented: "No guts, no glory. You have to get in there and mix it up with him!" BVR combat was all very well, but unless the entire enemy force could be taken out at medium range - a most unlikely assumption at supersonic speeds - the fight would quickly close to visual.

Maneuver combat demands a high thrust/weight ratio and a low wing loading. A delta wing was selected to give a high lift/drag ratio and a high coefficient of lift. It also has no clearly defined stall point, and gives low supersonic wave drag, but in its pure form lacks maneuverability. (For more information, see the section on the Mirage 2000.)

The Russians had largely overcome this fault by using a tailed delta configuration. A traditional horizontal tailplane provides a stabilizing download, but this reduces supersonic maneuverability. The new European fighter used canard foreplanes to overcome the control problems. Canards add to lift, giving a destabilizing effect. While this places exceptional demands on the active fly-by-wire flight control system, the use of canards made fully variable camber possible, by combining leading edge flaps with trailing edge flaperons.

EAP

The next step was to produce a technology demonstrator, the Experimental Aircraft Program (EAP), which was built by BAe and first flown in August 1986. EAP featured a cranked delta wing with automatic camber; quadruplex fly-by-wire; a chin inlet with a droopable lower lip to improve airflow at high alpha; and a huge vertical tail "borrowed" from Tornado. The canard foreplanes were set well forward to give a long moment arm for greater responsiveness. By default, it was powered by two Turbo-Union RB 199 augmented turbofans.

In the course of 259 sorties, EAP provided much valuable information. The maximum

BELOW: Afterburners blazing, the first Typhoon to be powered by EJ 200 turbofans soars into the sky. Almost certainly, thrust vectoring will be added, to improve the already impressive maneuverability still more.

coefficient of lift (CL$_{max}$) was found to change non-linearly, from totally unstable just before CL$_{max}$ was reached, to totally stable just afterwards. This was amended by straightening the wing leading edge to an angle of 53 degrees. Other, more minor changes, were put in hand.

FURTHER DEVELOPMENT

Stealth also became a consideration. The boxy chin intakes were curved upwards, to give a "smiling" appearance. Curved inlet ducts were used to mask the engine compressor faces, while RAM tiles and RAM paint were used to shield reflective surfaces. While the actual figure is classified, the result was to reduce the head-on RCS to between 20 and 25 percent of that of a conventional fighter (from 5m^2 to 1m^2). While this is of course worth having, in combat the advantage in early detection is fairly marginal - a matter of a few miles, or in time, a few seconds.

In 1992, disaster threatened. The dissolution of the Warsaw Pact and the collapse of the Soviet Union effectively removed the threat which the new European fighter was designed to counter. The new reunited German Republic queried the justification for the Eurofighter and sought cheaper alternatives. This was strenuously opposed by both Britain and Italy, and the project was eventually rescued, albeit at the cost of considerable bad feeling. Meanwhile Spain had acquired the F/A-18 Hornet as a stopgap, and there can be little doubt that if EF 2000 had been cancelled they would have bought more to fill the deficit.

The digital quadruplex fly-by-wire system had provided problems which took some considerable time to sort out. Automatic limits were built into the software to ensure that the aircraft would never depart controlled flight, nor could it be overstressed. This provided carefree handling, and reduced gust response, ironing out the "bumps" in high-speed, low-level flight.

Supersonic flight demands great lateral stability, while maneuver at high alpha can easily blanket the vertical tail. EF 2000 has a single tall fin for stability and controllability. It also has small horizontal fences just above the canard foreplanes and just below the cockpit. The purpose of these is to control the airflow over the fuselage at high alpha, but they also double as a step into the cockpit.

Low wing loading, essentially for maneuverability, is a compromise between structural weight and area. There is however a practical limit to area; once past this point weight saving becomes critical. This has been achieved by extensive use of carbon fiber composites (CFC). More than 40 percent of

the total structural weight is made up of CFC, as is about 70 percent of the surface area, including most of the wings, fuselage, and fin. Where birdstrike is a possibility, against the wing and fin leading edges for example, aluminum-lithium is used, while the canard foreplanes are of superplasticaly formed and diffusion-bonded titanium alloy.

ECR 90 has a total of 31 modes, with adaptive scanning, and is widely supposed to be able to handle up to ten simultaneous attacks. It is supplemented by IRST, and will have a helmet-mounted sight for off-boresight missile cueing. Defensively, EF 2000 is fitted with the Defensive Aids SubSystem (DASS), housed mainly in wingtip pods, although this is an area where savings have been made. Initially it included active radar jamming with highly directional phased array antennae, electronic support items, a missile launch warning system, and a laser warning receiver. Towed radar decoys, otherwise known as turds, are also carried.

PROGRESS

The prototype made its first flight from Manching in Germany on March 27 1994; the second from Warton in England ten days later. Both were powered by RB 199s, but the definitive EJ 200 powered all subsequent development aircraft. Typhoon, as it was named in September 1998, is scheduled to enter service with Britain and Italy in 2002. At some future date, it seems almost certain that thrust vectoring nozzles will be added to give post-stall maneuverability.

ABOVE: The traditional wingtip missile rails have been dropped in favor of pods housing ECM and other defensive aids systems. The actual fit varies with the customer.

France
DASSAULT MIRAGE 2000

ABOVE: Three huge drop tanks provide a long ferry range for the Mirage 2000-5, although they would have to be jettisoned for close combat.

The Mirage 2000 is unusual in that instead of being designed to meet an official specification the requirement was written to match the predicted performance of the airplane! Its history is tortuous.

The tail-less delta configuration was first used by the Dassault company in the mid-1950s. The threat was at that time the fast, high-flying jet bomber. The delta wing, with its combination of high sweepback and large area (with low wing loading), was optimum for the top right-hand corner of the flight performance envelope (maximum speed and maximum altitude). It had other advantages: the long chord at the root provided adequate depth for fuel tankage, and made construction simple.

There was of course a price to be paid. Maximum lift was reached at a relatively high alpha which could not be attained on take-off or landing without grounding the rear end. At these times, nose-up pitch control was effected by deflecting the trailing edge elevons upwards. This was less than satisfactory. On take-off it effectively reduced lift just when it was most needed. On landing, it had exactly the same effect of reducing

lift, which meant that a far greater speed margin was needed on the approach if unacceptably high sink rates were to be avoided. Either way, take-off and landing speeds were higher, and ground rolls much longer, than would have been the case with a more conventional layout. Finally, a wide margin of static stability was needed in all flight regimes, especially when carrying external stores.

There was also the problem of getting on the back of the drag curve. A delta wing has no clearly defined point of stall. Increased alpha gives increased lift, but also increased induced drag. The critical point comes when total drag exceeds the available thrust of the engine. Speed then starts to bleed off, and continues to do so until ultimately flying speed is lost.

The answer is to drop the nose and reduce alpha, thereby reducing drag to the point where an acceleration capability is regained. This involves loss of altitude, which is fine in most flight regimes, but not a good idea on final approach. To get on the back of the drag curve when landing is usually the prelude to disaster.

MIRAGE 2000-5

DIMENSIONS: Span 29ft 11in/ 9.13m; Length 48ft 1in/ 14.65m; Height 17ft 1in/5.20m; Wing Area 441sq.ft/41m²;

ASPECT RATIO 2.03

WEIGHTS: Empty 16,535lb/7,500kg; Takeoff c25,353lb/11,500kg

POWER 1xSNECMA M53-P2 turbofan; Maximum Thrust 21,400lb/9,707kg; Military Thrust; 14,400lb/6,350kg

FUEL: Internal 7,055lb/3,200kg; Fraction 0.28

LOADINGS: Thrust-max 0.84lb/lb-kg/kg; Wing 57lb/sq.ft-261kg/m²

PERFORMANCE: V_{max} high Mach 2.20 V_{max} low Mach 1.20; V_{min} 100kt/185kmh; Operational Ceiling 59,058ft/ 18,000m; Climb Rate 58,000ft/min-295m/sec

WEAPONRY: 2x30mm DEFA 554 cannon with 125rpg; 6 or 8 MICA AR or IR homers

USERS: Abu Dhabi, Egypt, France, Greece, India, Peru, Qatar, Taiwan.

ABOVE: The semi-circular intakes with translating shock cones are typical of all Dassault deltas. This is a two-seater 2000D.

FAR RIGHT: The Mirage 2000 was optimised for high speed, high altitude interception. This example carries four MICA and two R550 Magic AAMs.

By the same token, hard turns at high alpha bled off speed alarmingly. But, for defense against jet bombers, little maneuver was thought necessary, and the drawbacks of the tail-less delta were considered to be more than outweighed by the advantages.

The result was the Mirage III, which came close to the ideal of "the smallest possible airframe wrapped around the largest available engine!" Judged by the performance and maneuverability standards of its contemporaries, it was a very ordinary fighter. Then in the early 1960s, Israel, beset on all sides by hostile Arab nations, many of whom were equipping with the MiG-21, went shopping for something with which to oppose them. At that time, British and American fighters were embargoed, which left little choice. Israel had previously acquired fighters from Dassault; now they bought the Mirage IIICJ.

In the bitter wars that followed, the Mirage IIICJ gained an enviable reputation, its pilots consistently outfighting the formidable MiG-21. That pilot quality was the dominant factor was largely ignored; the French fighter had gained that most valuable of cachets, "combat proven!" For a few years, the Mirage III was the most famous fighter in the world, operated by more than 20 nations.

THE NEXT GENERATION
This enviable reputation notwithstanding, France's l'Armée de l'Air was fully aware of its shortcomings. To succeed it, Dassault reverted to a conventional swept-wing layout. The Mirage F.1 was of similar overall dimensions, but was significantly heavier and had a much smaller wing area than its predecessor. All else being equal, the greatly increased wing loading should have

adversely affected both minimum speeds and maneuver capability. In fact it was more than offset by a combination of flaps and slats. These reduced the approach speed by a massive 30 percent, reducing the length of runway needed. At most speed/altitude combinations the F.1 could pull one more g than the III, while much less energy was bled off in hard turns. The Mirage F.1 appeared to have sounded the death-knell of the tail-less delta.

After a brief flirtation with variable sweep, Dassault began work on a conventionally winged, twin-engined single-seater: the Super Mirage. Projected performance was to rival the F-15 Eagle, but Dassault, quickly realizing that this was unaffordable, began private studies in 1972 for a cheaper alternative. Once again they appraised the advantages of the tail-less delta, knowing that the latest technological advances, already explored for the Super Mirage, would be able to minimize the disadvantages of the configuration. When in December 1975 the Super Mirage was finally canceled, its replacement was already waiting in the wings.

THE DELTA REVIVED
As the Mirage 2000, the new fighter first flew on March 10 1978. Both in appearance and in dimensions, it was very similar to the Mirage III, although leading edge sweep was reduced from 60 to 58 degrees. Careful wing/body blending was used to minimize wave drag; it also increased body lift at high alpha. Like its Mirage III ancestor, the rear of its canopy was faired into the body, at a slight penalty in rearward visibility. Composites, both carbon fiber and boron, were used to save weight, although not as extensively as in later machines.

Under the skin it was a very different bird. Its wings carried full span slats to the leading edges, and full-span two-piece elevons to the trailing edges, to give automatic variable camber. High lift devices like these had proved impossible to use on the earlier machine; advances in computers now made them a practical proposition. These control surfaces could be configured to give extra lift in most flight regimes, rather than imposing downloads at times when they were most undesirable.

The key that really unlocked the door was moving the center of gravity aft to give relaxed stability, coupled with quadruplex fly-by-wire. With no wide range of static stability, there was little or no inherent inertia to be overcome when commencing a maneuver, which aided rapid control response and fast transients between flight modes. Finally,

RIGHT: The "glass cockpit" of the Mirage 2000-5 has four multi-function color displays. A first in a fighter was the "look level" display just beneath the HUD.

as with all FBW systems, alpha limits were set to give carefree handling.

To improve high alpha capability still more, small strakes were mounted on the outside of the intake ducts and slightly higher than the wing, which generated vortices to clean up the airflow over the wing surface.

SPEED AND ALTITUDE

At a time when most of the world's major aircraft producers were concentrating on agile tactical fighters, both Dassault and l'Armée de l'Air kept their sights firmly fixed on the top right hand corner of the flight performance envelope (max speed/max altitude), which was one of the main reasons for reverting to the tail-less delta. The engine selected was also unusual in this respect; it was the SNECMA M53. While almost every fighter engine of this and subsequent eras used two-spool turbofans, the M53 combined a single spool, with a low bypass ratio and a moderate compression ratio. In fact, it is more nearly a continuous bleed turbojet than a turbofan, and as such it is most efficient in the

high altitude supersonic regime than more conventional fighter turbofans. At subsonic speeds, specific fuel consumption is somewhat on the high side.

The M53 has been cleared for speeds up to Mach 2.5, although kinetic heating restricts the Mirage 2000 airframe to Mach 2.2. This could obviously be exceeded if the situation warranted it. It is fed by variable side inlets which are pure Dassault, with translating half shock cones known as "souris," or mice.

MANEUVER

Like most modern fighters, the Mirage 2000 is stressed for loadings of +9/-3g, although in extreme cicumstances the system can be overridden to allow +13.5g. Maximum roll rate is fast at 270deg/sec, and pitch rates appear to be rapid. Instantaneous turn rate varies between 20 and 30deg/sec, but sustained turn is, as one might expect from the relatively low thrust/weight ratio, a little lacking, as is acceleration. Unofficial figures give the following at 5,000ft (1,524m): 14deg/sec at Mach 0.5, increasing to 17deg/sec at Mach

0.9, then decreasing to 7deg/sec at Mach 1.2. Up at 30,000ft (9,144m), these become 7deg/sec at Mach 0.9, falling to 4deg/sec at Mach 1.6. It must be said, however, that these figures are in line with other modern fighters, both American and Russian. Controllability is stated to be adequate down to 40kt (74kmh) in a descending maneuver, although minimum speed for stable flight is 100kt (185kmh).

COCKPIT AND AVIONICS

The first Mirage 2000C entered service in November 1982, and since then several different variants, including the two-seater Mirage 2000N interdictor, have followed. The latest is the Mirage 2000-5 multi-role fighter, the first deliveries of which took place in 1996.

Dassault followed McDonnell Douglas in providing HOTAS and a "glass cockpit," with three CRT displays on the dash. While these are normally head-down displays, one of them was mounted just beneath the HUD to provide a "look-level" capability. Two of these presented radar and flight data, systems and weapons status, all of which could be transferred to the HUD at need, while the third displayed threat indications provided by the comprehensive EW suite.

In 1991, the Advanced Pilot/System Interface (APSI) cockpit was introduced to fighter construction, and this features on the Mirage 2000-5. It has five electronic displays. The Thomson-CSF wide angle HUD, with a field of view of 30deg x 20deg, surmounts the head-level display, which shows radar, forward-looking infra-red, or laser designator imagery, without the pilot having to look down. Below this, flanking the dash, are two multi-mode displays, and at the center bottom is the color synthetic tactical display.

The initial radar choice was, in all except a small early production batch, the Thomson-CSF RDI, which was optimized for the interception mission. For unambiguous range and velocity data in search mode against on distant targets, it used very high PRFs (pulse repetition frequencies). Maximum range was stated to be 65nm (120km) against head-on fighter-sized targets, reducing to 30nm (56km) against tail-on contacts or low flying aircraft in look-down mode. While this did not match contemporary American radars such as the Hughes APG-65, l'Armée de l'Air felt that, given the small RCS of the Mirage 2000, it was enough to give a decisive advantage over the threat radars of the time. It could also provide guidance for missiles to "snap-up" against very high flying opponents.

The latest radar is the Thomson-CSF RDY, with rather longer range and greater capabilities than its predecessor. It can detect up to 24 targets simultaneously, while tracking eight and prioritizing the greatest threats.

Radar-homing missiles usually announce their presence with electronic emissions, and the EW suite is programmed to respond automatically. But heat-homers are passive, and therefore much more dangerous. To counter these, Matra has developed SAMIR, an infra-red system to detect the heat signature of a missile during its motor burn time, giving an audible warning, and an indication of the attack quadrant. SAMIRs will be mounted at the rear of the underwing Magic missile launchers, giving all-round coverage, and will be coupled to the Spirale decoy system.

BELOW: Abu Dhabi is one of the export customers; the two nearest are single-seater Mirage 2000EAD; furthest away is the two-seater DAD.

France

DASSAULT RAFALE

In the late 1950s, Dassault had scaled up the Mirage III into the twin-engined, two-seater Mirage IV strike aircraft. Now, with the Mirage 2000 safely launched, Dassault traversed the same road to produce the Super Mirage 4000, a scaled-up, twin-engined and potentially more capable version of the earlier fighter. It differed in one important respect, however: it featured movable canard foreplanes on the intake ducts, thus predating the next generation of European canard deltas. It must of course be stated that Sweden's Saab JA 37 Viggen is excluded from this category as not only were its foreplanes fixed (with moving elevators), they were solely for short field performance, tended to stall before the wing, and added little or nothing to combat maneuverability.

While the Super Mirage 4000 undoubtedly had tremendous potential, it was overtaken by events. The American F-16 had set new standards for affordable fighter maneuverability, while the main threat, the Soviet Union, was also known to be developing very maneuverable fighters. The accent had switched to agility.

· As related in a previous section, the major European nations combined their resources to produce a new agile fighter of canard delta configuration, which eventually emerged as the EF 2000 Typhoon. Dassault was originally involved, but intransigence lost them the day. They demanded design leadership on the grounds that they had more experience of deltas (true); industrial leadership (on what grounds was uncertain); an unacceptably high proportion of production (national aggrandisement). Finally, their stated weight limit virtually guaranteed that only the French SNECMA M88 turbofan would be suitable. This was unacceptable to the other nations involved, and it also raised questions as to whether the radar and other systems might not go the same way under French design leadership. *Bonne chance* to Dassault for trying, but it didn't work, and they were left isolated.

ABOVE: Rafale B shows its unique intake, located beneath a bulged fuselage section which acts as a compression wedge.

RAFALE

DIMENSIONS: Span 35ft 5in/10.80m; Length 50ft 1in/15.27m; Height 17ft 6in/5.34m; Wing Area 492sq.ft/45.70m²

ASPECT RATIO: 2.55.

WEIGHTS: Empty 19,974lb/9,060kg; Takeoff c32,430lb/14,710kg

POWER: 2xSNECMA M88-2 turbofans; Maximum Thrust 16,843lb/7,640kg; Military Thrust 11,243lb/5,100kg

FUEL: Internal 9,921lb/4,500kg Fraction 0.31

LOADINGS: Thrust -max 1.04lb/lb-kg/kg; Wing 68lb/sq.ft-322kg/m²

PERFORMANCE: V_{max} high Mach 1.80 plus; V_{max} low Mach 1.14; V_{min} 80kt/148kmh; Operational Ceiling 55,000ft/16,763m; Climb Rate 60,000ft/min-1,000m/sec

WEAPONRY: 1x30mm DEFA 719B cannon; 8xMICA AR and IR homers.

USERS: France

ABOVE: Although conceived as a fighter, Rafale B has become a two-seater attack aircraft with secondary air combat capability, as well as a conversion trainer.

BELOW: Rafale M, a single-seat carrier fighter is stated to be "good around the boat" although the approach angle looks a bit high in this picture.

Dassault then proceeded alone with the Avion de Combat Experimentale (ACX), later to become Rafale. As with the EF 2000, a technology demonstrator was built. This was dimensionally slightly larger than the definitive aircraft, and, pending the development of the new SNECMA M88-2, was powered by two General Electric F404 augmented turbofans. First flown on July 4 1986, it made its first public appearance as Rafale A at Farnborough Air Show, England, in September of that year. At its next appearance two years later, it demonstrated an impressive sustained turn rate, albeit at low speed and low level, of 24deg/sec. Gallic hubris finally went over the top at Farnborough '94, when after each flight the number of sorties was repainted on Rafale's tail to emphasize how far ahead they were of EF 2000!

Initial plans called for three main variants: the single-seater Rafale C fighter with a secondary attack capability; the two-seat Rafale B/D as either a conversion trainer or an attack aircraft with a secondary air superiority role; and Rafale M, a dedicated single-seat carrier fighter/attack aircraft.

Rafale C was to be the primary air superiority fighter for l'Armée de l'Air, but the dissolution of the Warsaw Pact and the break-up of the Soviet Union effectively removed the perceived threat. This caused a major rethink. A priority was to replace the elderly Jaguar attack aircraft, and this mission now assumed greater importance. Pilot workload was high, and the operational experience of French Jaguars in the Gulf War of 1991 clearly demonstrated the advantages of a two-man crew. The result was that *l'Armée de l'Air* requirements were drastically revised to include a majority of two-seat aircraft..

Economic considerations have delayed French orders for Rafale, and the *Aeronavale* will receive just 12 between 1999 and 2003, in which year deliveries to *l'Armée de l'Air* are scheduled to start. To date export orders have failed to materialize, although the Mirage 2000 is still selling. It seems probable that nations able to afford Rafale will go for something better.

AERODYNAMICS AND STRUCTURE
The finished aircraft is an attractive-looking fighter, its delta wing closely approximating that of the Mirage 2000, with leading edge extensions at the roots. The tips have rails for Magic 2 missiles. The canard foreplanes are set just above the inlets, where they impinge least on the pilot's view "out of the window," but they are close-coupled, with a short moment arm, and have correspondingly less authority than those of Typhoon.

Where Rafale really differs is the intakes themselves. An early design study featured a chin intake of the type seen on the F-16, but this was a complete departure from the semi-circular inlets previously favored by Dassault, and in any case would have involved a deeper fuselage. The solution adopted was to combine the best features of both, with the inlets (plain, as Mach 2 capability was not a requirement) located under a fuselage bulge which act as a forebody to smooth airflow at high alpha, much like the strakes of the F/A-18. The canard foreplanes are located on these bulges.

Much of the wing structure and fuselage cladding is of CFC, with Kevlar at the extreme rear, while the canards and leading edge slats are of titanium alloy. Aluminum lithium is used on part of the fuselage, and on the leading edge of the single fin. Propulsion

Unlike the M53, the M88 used to power Rafale is a genuine twin-spool turbofan. While the -2 motor is used for preproduction Rafales, the addition of a modified two-stage compressor and a low pressure turbine should increase projected maximum thrust to 23,600lb (10,705kg), increasing thrust loading to 1.46, remarkable in fighter configuration. This will of course be rather less for the navalized Rafale M. Stressed for carrier landings and catapult launches, with beefed-up landing gear, arrester hook, and nosewheel strut, it is significantly heavier than the land-based Rafale.

ABOVE: Rafale M armed with R550 Magic heat homers on wingtip rails, and four Matra MICA AAMs under the wings. Maximum AAM load is stated as eight MICAs.

PERFORMANCE AND HANDLING

Although the maximum brochure speed of Rafale is stated to be Mach 1.8, preproduction models have in fact exceeded Mach 2. Supercruise has also been demonstrated, although the actual speed attained has not been released. Roll rate is a sparkling 270deg/sec, while instantaneous turn exceeds 30deg/sec, and Rafale is stressed for loadings of +9/-3.2g. For carrier operations, landing alpha is somewhat high at 16-18 degrees, but this is not considered a problem by the *Aeronavale*.

COCKPIT AND AVIONICS

Rafale has a "glass cockpit," with a wide angle holographic HUD and three color multi-function displays, one of which is head-level as in the Mirage 2000-5. These are backed by a helmet-mounted sight which can be used to acquire and designate off-boresight targets. When the pilot has his hands full with HOTAS, voice control can be used. A vocabulary of 37 words, surprisingly in English, is projected. The seat is raked at an angle of 29deg to provide greater g-tolerance, while a sidestick controller has replaced the control column.

Radar is the multi-mode *Radar à Balayage Electronique* (RBE) 2. This is the first European fighter radar to use a phased array antenna. The antenna is fixed, and the actual scanning is steered electronically. It can operate in more than one mode at a time, changing from one to the other many times per second, and has far more potential and flexibility than trainable antennas. Range is rather greater than that of the Mirage 2000, 54nm (100km) even in look-down mode. Tracking parameters are probably similar, and at least four targets can be engaged simultaneously.

The RBE2 is backed by the *Optronique Secteur Frontale* (OSF), consisting of IRST, FLIR, and laser rangefinding. Maximum range of OSF in ideal climatic conditions is 43nm (80km). A full EW suite is carried, including the missile launch detector as described for the Mirage 2000.

STEALTH

Low-observable features have been incorporated into Rafale, which in any case has a fairly low basic RCS. The fin/fuselage junction was modified, RAM was applied, and the canopy was gold-filmed. But, as in Typhoon, these stealth measures will probably only reduce Rafale's RCS to about one-fifth.

Great Britain/USA
BAᴇ/BOEING HARRIER II PLUS

LEFT: The only really successful fast jet capable of operating without benefit of conventional runways is the Harrier. This is the British Sea Harrier FRS 2.

As noted for Tornado, it is far more difficult to turn an attack airplane into a fighter than vice versa. While fighter variants of the Harrier have short-comings in the air superiority arena, the type has one supreme advantage which makes its use worthwhile. This is the ability to take off and land vertically (VTOL), although to improve payload a short rolling take-off is almost always used. VTOL has now become STOVL (short take-off, vertical landing).

The Harrier can operate from areas too small for any other fighter to consider; for example, an urban car park. Whereas conventional fighters land and then stop, the Harrier's ability to stop and then land makes it uniquely suitable for forward basing. In naval service, not only can it operate from quite small ships, it can come aboard in quite appalling weather using the simple expedient of slowing right up, which enables its pilot to establish visual contact with little fear of either collision or missing the landing target alto-gether. In fact it can land in such conditions that would have the pilot of a conventional fast jet reaching for his bang-seat handle.

The advantages of VTOL were always obvi-ous, but attaining it was a long and hard road, with many dead ends. There was no great problem in getting an aircraft to take off verti-cally; all it needed was a sufficient margin of thrust over weight. Control in the hover, or at low speeds, was another matter. Sufficient high pressure air had to be bled off to supply a reaction control system for pitch, roll, and turn.

AV-8B HARRIER II PLUS

DIMENSIONS: Span 30ft 4in/9.25m; Length 47ft 9in/14.55m; Height 11ft 8in/ 3.56m; Wing Area 230sq.ft/21.37m^2

ASPECT RATIO: 4.0

WEIGHTS: Empty 14,867lb/ 6,744kg; Takeoff c23,650lb/10,728kg

POWER: 1xF400-RR-408 unaugmented turbofan; Military Thrust 23,800lb/10,796kg

FUEL: Internal 7,561lb/ 3,430kg; Fraction 0.32

LOADINGS: Thrust-mil 1.01lb/lb-kg/kg; Wing 103lb/sq.ft-502kg/m^2

PERFORMANCE: V_{max} high Mach 0.91; V_{max} low Mach 0.86; V_{min} Zero; Operational Ceiling c45,000ft/13,715m; Climb Rate c50,000ft/min-254m/sec

WEAPONRY: 1x25mm GAU-12 cannon with 300 rounds; 6xAIM-120 Amraam AR missiles or 2xAmraam and four Sidewinders

USERS: USA, Italy, Spain.

Most early experiments involved tail-sitters. Only the power source varied, from large turboshafts with huge contra-rotating propellers, to the tiny Ryan X-13 Vertijet which pioneered jet lift. But tail-sitters posed almost insuperable problems. Access for maintenance and servicing was difficult, while entering the cockpit of a tail-sitter was a job for a contortionist. But the real problem was that the decelerating transition from wingborne flight to the vertical hover, and the subsequent landing, made backwards with the pilot's feet higher than his head, all the while maintaining a true vertical descent, in crosswinds or in reduced visibility, demanded a standard of flying skill which was far beyond the average.

This left the flat riser, which could if necessary take off and land conventionally, as the only viable option. The simple solution appeared to be a battery of small and simple lift engines allied to a large propulsion unit. The nearest this ever got to a fighter was in France, where Dassault produced the Mirage IIIV, only to find that the weight and volume

penalties of eight lift engines clustered around the aircraft center of gravity were unacceptable.

The next step was thrust vectoring plus lift engines. The German experimental VJ-101C combined two lift engines mounted in the fuselage with a pair of engines on each wingtip in swiveling pods, to provide either forward or vertical thrust. This machine exceeded Mach 1 on several occasions, but for many reasons could never have been developed into an operational aircraft. The only success in this field was Russian. The Yak-38 combined two lift engines just aft of the cockpit with a single propulsion engine with pivoting nozzles on a bifurcated duct.

One of the greatest problems was balancing the thrust of multiple engines. The ideal solution was to have a single engine producing thrust through swiveling nozzles spaced around the center of gravity. The idea came from Frenchman Michel Wibault, but was developed and turned into a practical proposition by the British companies Bristol and Hawker. The heart of the project was the Bristol Pegasus turbofan, with two vectoring nozzles supplied by "cool" air (although you wouldn't put your finger in it) from the compressor, and two more vectoring nozzles at the hot end, fed by a bifurcated trunk. Hawker then designed a suitable airframe around it. The result was the P.1127, which first flew in November 1960.

The P.1127 was marginal in every way. It was therefore hardly surprising that progress was slow. But gradually more thrust was wrung from the Pegasus; the airframe was refined; and capability increased. At last a usable warplane emerged. The first production Harrier, a close air support aircraft, made its maiden flight on December 28 1967.

The Harrier showed little promise as a fighter. Its huge side intakes were sized for maximum thrust at low forward speed; as the compressor had to feed the front nozzles this was unavoidable, but it ruled out supersonic flight, let alone Mach 2. In any case, the Pegasus had no afterburner. The high wing loading meant poor turning capability, and the nose was on the small side to house a really capable radar.

HARRIER FIGHTERS

In 1966, the British government of the day decided that air defense of the fleet could be left to the RAF, itself a diminishing and over-stretched organization. By the time common sense prevailed, and the need for organic air power at sea was acknowledged, the only air platforms the Royal Navy had were helicopter carriers, officially known as "through-

BELOW: Cockpit displays of the Harrier II Plus, with targets displayed in the HUD. On the right is the air picture; on the left the ground mapping display.

deck cruisers." As these were far too small to operate conventional machines, the Harrier was the only possible choice.

The first fighter variant was the Sea Harrier FRS. 1. First flown on August 20 1978, it gained the cachet of "combat proven" in the South Atlantic War of 1982, where, armed with the latest AIM-9L Sidewinders, it took on and defeated supersonic Argentinian Mirage III fighters. It was of course pilot quality that won the day, rather than the intrinsic worth of the Sea Harrier, but it had proved one thing. Fears had been expressed about the vulnerability of the reaction control system to battle damage; these were finally laid to rest.

However, many worthwhile lessons were learned. The Sea Harrier FRS.2, which first flew on September 19 1989, carried a far better, multi-mode pulse-Doppler radar, and provision was made for at least four AAMs rather than two, with some BVR capability. Had it not been for the fact that most FRS.2s were to be rebuilt FRS.1s, it could also have had a bigger wing.

US MARINES SERVICE
The US Marines had operated Harriers from a very early stage. Built under license by McDonnell Douglas (as the company then was) with Pratt & Whitney producing the Pegasus engine as the F400-RR, the AV-8A, as the Harrier became in USMC service, was an instant success. It filled their greatest need: rapid reaction close air support, for which it could be based on small helicopter carriers, or commando assault ships close inshore. Then, when the battle moved inland, the AV-8s could follow, using ad hoc bases.

The AV-8A was followed in USMC service by the AV-8B Harrier II, which a combination of minor airframe changes and a larger wing achieved double payload/range performance, for a small reduction in maximum speed. Weight was held down by extensive use of composites; digital fly-by-wire was incorporated; and the avionics suite was upgraded.

The USMC had been among the first to experiment with Vectoring in Forward Flight (VIFF), which allowed Harriers to perform some rather unorthodox maneuvers. Now they decided to go the distance and produce a dedicated fighter version. Ordered in 1990, this duly emerged as the AV-8B Harrier II Plus. Externally it is almost identical to the AV-8B, but has a slightly bulged radome to house the Hughes APG-65 multi-mode radar as used in the Hornet. The cockpit is small, as in all Harriers, and space for visual aids is limited, with only two color multi-function displays and HUD. But, whatever its shortcomings, the Harrier II Plus gives the USMC a credible quick-reaction air defense capability.

ABOVE: Loaded for bear, two USMC Harrier II Plus are outbound in the air superiority missions. Although not supersonic, the Harrier II Plus gives the USMC a credible air defense capability.

Russia
MAPO MiG-29 FULCRUM

ABOVE: Initially, one of the greatest Russian threats to NATO aircraft, on the unification of Germany, former East German MiG-29s were transferred to the *Luftwaffe* with which ironically, the MiG-29 now flies as part of that organization. This MiG-29 is assigned to JG 73.

Spurred by the American trend towards agile fighters, the Mikoyan OKB commenced work on a counter in the early 1970s. The baseline used was research done by the Central Institute of Aero-Hydrodynamics (TsAGI), which strongly influenced the layout. The result was the MiG-29, given the NATO reporting name of Fulcrum, which was first flown by Alexsandr Fedotov on October 6 1977, and which started to enter Russian service in June 1983.

FULCRUM REVEALED
For many years, Fulcrum was surrounded by the traditional Soviet veil of secrecy, which was finally lifted in September 1988, when two MiG-29s were displayed at Farnborough Air Show, England. In many ways the aircraft is conventional enough, with a wing leading edge sweep of 40 degrees and large and thick leading edge root extensions (LERX), which act as compression wedges at high alpha; automatic slats and flaps; twin fins and rudders canted outwards at six degrees; a hydraulic flight control system; and a conventional cockpit. In other ways it is radically different.

The augmented turbofans are widely spaced, with a Tomcat-like pancake aft fuselage. This aspect initially led to speculation that the engines were sensitive to disturbed airflow. Nothing could have been further from the truth. The rectangular intakes are of variable geometry to allow speeds up to Mach 2.3. But they contain one unique feature: the top-hinged ramp can be lowered to blank off the intake almost completely, with the engines fed through louvers on the top of the LERX. This is primarily a measure to prevent stones and slush from entering the intakes on take-off. Under normal circumstances the inlets close automatically when speed drops to 108kt (200kmh), but they can also be selected

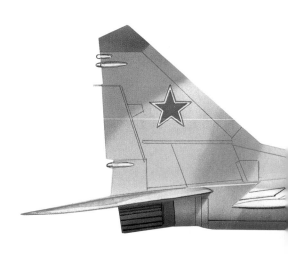

MiG-29M

DIMENSIONS: Span 37ft 3in/11.36m; Length 57ft 0in/17.37m; Height 15ft 6in/4.73m; Wing Area 409sq.ft/38m^2

ASPECT RATIO: 3.40

WEIGHTS: Empty: c24,250lb/11,000kg; Takeoff 37,038lb/16,800kg

POWER: 2xKlimov RD-33K; Maximum thrust 19,400lb/8,800kg; Military thrust 12,125lb/5,500kg

FUEL: Internal 9,978lb/4,526kg; Fraction 0.27

LOADINGS: Thrust (max) 1.05; Wing 91lb/sq.ft-442kg/m^2

PERFORMANCE V$_{max}$ high Mach 2.30; V$_{max}$ low Mach 1.23; V$_{min}$ c110kt/204kmh; Operational Ceiling 55,777ft/17,000m; Climb Rate 64,945ft/min-330m/sec

WEAPONRY: 1x30mm GsH 301 cannon with 100 rounds; 4xRVV-AE AR homers and four R-77 IR homers.

USERS: (all variants) Belarus, Bulgaria, Cuba, Czech Republic, Germany, Hungary, India, Iran, Iraq, Kazakhstan, Malaysia, Moldova, North Korea, Poland, Romania, Russia, Serbia, Slovakia, Syria, Turkmenistan, Ukraine, USA, Uzbekistan, Yemen.

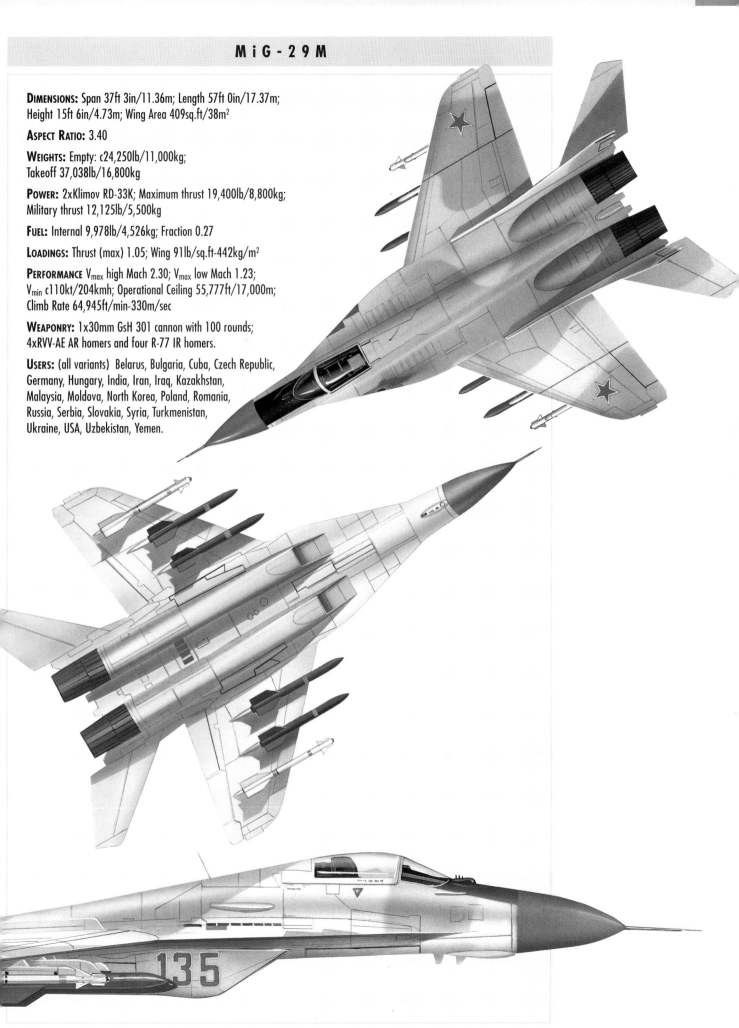

ABOVE: The widely spaced engines and leading edge root extensions are clearly visible in this plan view of the MiG-29, seen at Farnborough Air Show, 1988.

BELOW: MiG test pilot Anatoliy Kvotchur "demonstrates" the K-36D ejection seat at Le Bourget 1989. His MiG-29 is about to hit the ground, while Kvotchur, his parachute only half-deployed, is seen at left mere seconds from touchdown. Way above him is the abandoned cockpit canopy.

closed in flight, with a limiting speed of 432kt (800kmh). Obviously a useful precaution against birdstrike at low level, this also demonstrates that the engines are remarkably tolerant of disturbed airflow.

The widely spaced engine nacelles provide extra keel area, which adds to longitudinal stability at supersonic speeds, but also has one disadvantage. The main landing gears had to be located outboard, retracting forward into the wing roots. To avoid excessive length and complexity, ground clearance is restricted. This was almost certainly a factor in the design of the intake blanking arrangement, as it would be all too easy for the nosewheel to throw up debris which could then be ingested.

Metal honeycomb is used for all control surfaces, while composites and fiberglass are used for much of the skinning. External finish is typical Russian "good enough," with ill-fitting access panels. Under the surface, titanium alloy is widely used for structural members.

Previous Russian fighters had been noted for the poor view "out of the window." Fulcrum's design went a long way towards redressing this, with a bubble canopy and a wrap-around windshield although, like Typhoon and Rafale, the canopy was faired into the rear fuselage. A few switches were located on the control column, but these were far from providing a full HOTAS capability. However, one area of excellence was provided by the K-36D ejection seat, the capabilities of which were inadvertently demonstrated by Anatoly Kvotchur at the Paris Air Show in 1989. Having lost control following a birdstrike, he ejected just outside

the safe limits of the seat and was deposited on the ground with only a black eye to show for the experience.

AVIONICS

Although the previous generation technology of the early MiG-29 caused it to be dubbed "the last of the dinosaurs" by at least one commentator, it had one feature in which it clearly outstripped the West. This was the SUV fire control system. This combined the N 019/RP-29 Sapfir multi-mode pulse-Doppler radar linked to a laser rangefinder, an infra-red search and tracking (IRST) system in a domed housing slightly offset in front of the cockpit, and a helmet-mounted sight which allowed missile launches at high off-boresight targets.

The Sapfir-29 is stated to have a detection range of 54nm (100km) against a fighter-sized target, and a tracking range of 38nm (70km). Ranges for look-down mode were not stated, but are probably about two-thirds of these figures. It can track up to ten contacts simultaneously but has no multiple target engagement capability. For gun tracking, the IRST coupled with the laser rangefinder give unparalleled accuracy. In trials, targets were often destroyed with bursts of between three and five rounds. If the target disappeared into cloud, radar tracking automatically took over when the IRST lost contact.

HANDLING

The MiG-29 is believed to be virtually stall- and spin-proof. At 216kt (400kmh) its sustained turn radius is just 738ft (225m), while at double this speed it becomes 1,148ft (350m). Linear acceleration at sea level is 36ft/sec^2 (11m/sec^2), falling to just under two thirds of this figure at 19,686ft (6,000m). An automatic limiter restricts alpha to 26 degrees, although this can be exceeded.

CONVERSION TRAINER

Like most modern jet fighters, Fulcrum has a two-seater conversion trainer, the MiG-29UB. In keeping with standard Russian practice, the second seat is located ahead of the main cockpit, thus occupying much of the space used for radar and avionics. Consequently the MiG-29UB has no combat capability. It also differs from the single-seater in that the twin fins have no leading edge extensions, which are otherwise used to house IR flares and other countermeasures.

VARIANTS AND UPGRADES

The greatest shortcoming of the basic MiG-29 was endurance. To overcome this, the MiG-29S was developed, and was first flown on

December 23 1980. Externally it differs by having a bulged spine behind the cockpit to hold more fuel and additional avionics. Other differences are five-section leading edge slats, as against four previously; alpha limits increased to 28 degrees; the ability to engage two targets simultaneously; and compatibility with the RVV-AE "Amraamski" AAM. Radar is the improved N 019M. The export version, the MiG-29SE, differs only in having a slightly less capable radar.

While the MiG-29M looks very much a standard Fulcrum, under the skin it is a very different bird. First flown on April 25 1986 by Valery Menitsky, it is powered by uprated RD-33K turbofans, on which the inlet blanks are replaced by deflector grilles, with the intake louvers on the LERX deleted. Major structural changes include a redesigned wing, with a sharp leading edge and extended ailerons, which also contain greater fuel capacity. The nose is slightly lengthened and the dorsal spine enlarged to increase internal fuel capacity. The horizontal tail surfaces have been enlarged and have a notched leading edge.

Quadruplex fly-by-wire has replaced the hydraulic flight control system, and the avionics have been updated. Radar is now the multimode Phazotron N 010 Zhuk, with a 26.77in (68cm) diameter antenna, able to handle up to four simultaneous attacks, and with detection and tracking ranges believed to significantly exceed those of the Sapfir-29. IRST and helmet-mounted designator are also believed to have been considerably improved. The pilot's seat is raised; two MFDs dominate the dash; and HOTAS has been incorporated.

The need for a carrier fighter led to the MiG-29K development project. Based on the MiG-29M, this featured structural strengthening, a beefed-up landing gear, folding wings, and a tail-hook. Wing area was also increased a little. As Russian naval ambitions appear to have subsided, this project is almost certainly now defunct.

Myths and legends surround the MiG-29. It has been widely reported to be superior to the F-15 in BVR combat, and superior to the F-16 in close combat. But in the Gulf War of 1991, in Iraqi service, it achieved nothing.

BELOW: The MiG-29SM, seen here fully loaded for air combat. Whatever the myths surrounding this fighter, it has achieved nothing in this role so far. Involved in the Gulf War of 1991, NATO action against Serbia in 1999, and the fracas between Sudan and Ethiopia in the same year, it has suffered many losses but scored no air combat victories.

Russia

MAPO MiG-31 FOXHOUND

Towards the end of the Cold War, the strategic air threat to the Soviet Union changed. It became a combination of cruise missiles launched from outside Soviet air space, and the low level B-1B penetrator. This posed new problems. The small and sleek cruise missiles could be launched in salvoes to swamp the Soviet defenses. They could come in at high or low level, and had a small RCS which made them difficult to detect and to attack. The B-1B, although not exactly a stealth airplane, had a relatively small RCS for its physical size, plus a comprehensive ECM suite. It could also carry cruise missiles.

To defend against this formidable multiple threat, a new fighter was needed. The Russians, long noted for screwing every last ounce of capability from existing aircraft, took the airframe of the legendary MiG-25, NATO reporting name Foxbat, and modified it to give a whole new range of capabilities.

Foxbat had originally been developed to counter the high-flying, trisonic Lockheed A-12, ancestor of the famous SR-71 Blackbird reconnaissance aircraft. It was designed for highest performance envelope: trisonic speed, ultra-high altitude, and an exceptional rate of climb. As there was little profit to be gained from a tailchase, interception was to be made from the front quarter.

Given closing speeds approaching one nautical mile (1.85km) per second and turn radii measured in dozens of miles, this made interception an extremely precise affair. Even the most gifted pilot was unable to cope with the split-second timing involved, and semi-automated interception was adopted, in which Foxbat was controlled by signals from the ground tracking stations which were fed into the automatic flight control system via data link.

A Foxbat pilot was therefore responsible for take-off; the terminal phase of the interception, including missile launch; and the return to base and landing. For the initial interception he became a systems manager, monitoring events, fuel status, and engine temperatures

ABOVE: The MiG-31 Foxhound, seen here carrying four R-33 active radar homing missiles (NATO code-name AA-9 Amos), was for years the Russian primary air defense interceptor.

MiG-31 FOXHOUND

DIMENSIONS: Span 44ft 2in/13.46m; Length o/a 74ft 5in/22.69m; Height 20ft 2in/6.15m; Wing Area 663sq.ft/61.6m²

ASPECT RATIO: 2.94

WEIGHTS: Empty 48,104lb/21,820kg; Takeoff 90,389lb/41,000kg

FUEL: Internal (T-6) 36,045lb/16,350kg; Fraction 0.40

POWER 2xSoloviev D-30F6 afterburning bypass turbojets; Maximum Thrust 38,580lb/17,500kg; Military Thrust 20,944lb/9,500kg

LOADINGS: Thrust (max) 0.85lb/lb-kg/kg; Wing 136lb/sq.ft-666kg/m²

PERFORMANCE: V_{max} high Mach 2.83; V_{max} low Mach 1.23; V_{min} c140kt/260kmh; Operational Ceiling 67,589ft/20,600m; Climb Rate c41,000ft/min-208m/sec

WEAPONRY: 1x23mm GSh-6-23 cannon with 260 rounds; 4xR-33 AR homers; 2xR-60 IR homers

USERS: Russia.

ABOVE: Unlike the US Navy F-14
Tomcat, the back-seater in the
MiG-31 was largely enclosed, and
could take little or no part in
visual range combat. As Foxhound
was structurally load-limited,
this made little difference.

and revolutions. Foxbat was officially red-lined at Mach 2.83, but in dire need it could be pushed up to Mach 3.2, although this could ruin the engines by overspeeding, making a safe recovery problematical.

With the correct intercept geometry established, target acquisition was made via the huge Smertch-A radar. Effective range of this was barely 27nm (50km), giving perhaps half a minute or so before the merge - little enough time to lock on and launch the big R-40 SARH missiles carried on underwing pylons.

First flown in 1964, development problems were extreme and it was 1973 before Foxbat entered service. It was gradually upgraded, with more efficient engines, a pulse-Doppler radar with at least some look-down, shoot-down capability, and more AAMs. By 1979 all interceptor Foxbats were of MiG-25PD standard, but still suffered from many of the original limitations.

Even before this, the cruise missile/B-1B threat loomed. To counter it, a completely different approach was necessary. The semi-automatic interception system remained effective at high altitude but was of little use against low-flying intruders. What was needed was an autonomous method of operation, free of the constraints of close ground control.

Basic requirements were enhanced endurance, and a superior avionics system able to detect and destroy multiple intruders at both high and low altitudes, with simultaneous attacks on multiple targets. This in turn demanded a dedicated weapons system operator. Finally, supersonic speed at low altitude, something the MiG-25 signally lacked, was a must.

REDESIGN

The starting point for the aircraft which was to become the MiG-31 was the Ye-155MP, first flown on September 16 1975 by Alexsandr Fedotov. Externally, it differed little from its MiG-25 ancestor. The fuselage was stretched to allow a second cockpit to be inserted behind the pilot, the canopy of which, with vision panels on each side, was faired into the dorsal spine. Speed brakes were located under each air intake duct; flaps and ailerons occupied the entire trailing edge of the wing; and the twin fins were given a slight increase in height.

The external similarity to the MiG-25 was largely illusory. Foxbat had been designed to soak in high kinetic heating levels at Mach 3 for extended periods, which had ruled out the use of aluminum in all but a few non-critical areas. Titanium was available, but at that time it was notoriously difficult to weld or to form. The only realistic alternative was nickel steel. The attendant weight penalty was minimized by making components as thin as possible, even though this limited structural loading while supersonic to 4.5g. But, as maneuver combat was never really an option, this hardly mattered.

By the gestation period of the MiG-31, technology allowed a different approach. Compared with Foxbat, steel content was reduced from 80 to 50 percent, titanium doubled to 16 percent and aluminum tripled to 33 percent. But weight savings were quickly eaten up by other requirements, although the supersonic loading limit was increased marginally to 5g.

The next four years saw other changes. Four-section slats and root extensions appeared on the wing leading edges and a retractable flight refueling probe was installed. Longitudinally staggered wheels on the main gears allowed taxiing on grass without making deep ruts.

Supersonic speed at low level was achieved by adding a third main wing spar to provide extra stiffening, plus a few structural tucks and gussets. Greater endurance was more complex. Internal fuel capacity on late model Foxbats was a massive 16 tons; for the MiG-31 this was increased still more, partly by using "wet" fins. The other need was for less thirsty engines.

The choice fell on the Soloviev D-30F6, described in the brochure as a bypass turbojet which, like its predecessors, uses high-density T-6 fuel, and has a hydraulically operated clamshell afterburner. Unusually, this uses fuel as the hydraulic fluid. Considerable structural alteration was needed to accommodate the D-30F6, which is limited to Mach 2.83.

The most notable feature of Foxhound was the Zaslon S-800 radar, which used an electronically steered phased array antenna, a first in the fighter world. Details are conflicting, but one Russian source states effective range as 108nm (200km), reducing to 65nm (120km) in look-down mode against fighter sized targets. Electronic scanning is supposed to give limited coverage astern. It is also stated to be effective against stealth aircraft, but what stealth aircraft, at what ranges, remains a mystery. What is certain is that it can track ten contacts, and guide four missiles at widely spaced targets simultaneously. The radar is backed by a retractable IRST under the nose.

Foxhound entered service in 1983. Although like Foxbat it can still operate under close ground control, current procedure is for four Foxhounds to be spaced 108nm (200km) apart and tied together by secure data link, to scan an area up to 486nm(900km) wide.

It was followed off the production lines in 1995 by the MiG-31M. This differs from the earlier version in having a reshaped cockpit with a single-piece wrap-around windshield; a deeper and broader dorsal spine which houses extra fuel; wingtip fairings for ECM; an improved radar which can handle up to six simultaneous attacks; more powerful turbojets; larger wing root extensions; and digital flight controls. The circular tactical display in the rear cockpit, which has just two small vision ports in the canopy, has been replaced by three MFDs. While the MiG-31M will not enter service as such, earlier Foxhounds may well be upgraded to this standard.

BELOW: Foxhound making first appearance in the West on its international debut, at Le Bourget, Paris, showing its flight refueling probe. Test pilot on the day was Roman Taskaev, who has since flown Foxhound over the North Pole on trials.

Russia
SUKHOI Su-27, -35 AND -37 FLANKER

On May 20 1977 the Sukhoi T-10-1 fighter prototype thundered down the runway at Zhukovsky and lifted into the air. At the controls was the diminutive figure of Vladimir Ilyushin, chief test pilot, former record breaker, and Hero of the Soviet Union.

For three decades Russian fighter design had been dominated by Mikoyan, to the point where MiG was synonymous with Russian fighter. Now things were about to change.

ORIGINS
The Soviet equivalent of the USAF "hi-lo" mix of fighters was paralleled by the Su-27 and the MiG-29. The influence of TsAGI on both fighters was apparent. The basic configurations had much in common: two widely spaced engines; high (by Western standards) aspect ratio wings; and twin fins. Common solutions to common problems! The T-10-1 (later Su-27, NATO reporting name Flanker) was designed to offset the F-15 Eagle, with comparable agility, long range and extended

combat persistence, while the MiG-29 was intended for the tactical arena.

All fighters are necessarily a compromise. Performance and agility are easily understood, but combat persistence is less so. It is a combination of flight endurance and on-board skills. A fighter which has gone "Winchester" (i.e., expended its air to air ordnance), may still have a bag full of fuel, but it no longer has combat persistence. Conversely, a fighter with weapons unexpended but which has reached "bingo" fuel (i.e., the minimum necessary to get home) must break contact with the enemy.

Combat persistence is difficult to achieve. The larger American fighters of the period - the Phantom, Tomcat, and Eagle - could carry up to eight AAMs apiece. For endurance they relied on a combination of external tanks (jugs) and flight refueling. Both have severe tactical limitations.

Flight refueling tankers are extremely vulnerable, and replenishment can be carried out only

ABOVE: Thrust vectoring allows the huge Su-37, seen here, to perform maneuvers impossible for conventional fighters, even though the nozzles move in the vertical plane only.

Su-35 SUPER FLANKER

DIMENSIONS: Span 48ft 3in/14.70m; Length 72ft 6in/22.10m; Height 20ft 9in/6.32m; Wing Area 668sq.ft/62.04m²

ASPECT RATIO: 3.48

WEIGHTS: Empty 40,565lb/18,400kg; Takeoff 56,658lb/25,700kg

FUEL: Internal c16,535lb/ 7,500kg; Fraction 0.29

POWER: 2xSaturn AL-31FM augmented turbofans; Maximum Thrust 30,865lb/14,000kg; Military Thrust c17,637lb/ 8,000kg

LOADINGS: Thrust 1.09lb/lb-kg/kg; Wing 85lb/sq.ft-414kg/m²

PERFORMANCE: V_{max} high Mach 2.35; V_{max} low Mach 1.14; V_{min} n/a; Operational Ceiling 59,058ft/18,000m; Climb Rate 60,000ft/min-305m/sec

WEAPONRY: 1x30mm GSh-301 cannon with 150 rounds; 14 AAMs; R-37, R-73; R-77 only on outboard stations.

USERS: (all variants) Belarus, China, India, Kazakhstan, Russia, Ukraine, Vietnam.

ABOVE: Flaps and gear down and dorsal speed brake extended, Sukhoi test pilot Eugeny Frolov brings the Su-37 in to land after a superlative aerobatic display.

at medium to high altitudes, where they are often in full view of enemy radar, and well back from any chance of attack from hostile fighters or SAMs. Buddy tanking is an alternative, but this effectively halves the number of tactical aircraft available for a mission, and with it, halves the munitions load carried.

Jugs cause extra drag, which reduces performance; as a rule of thumb, half the fuel carried in an external tank is used in getting the other half to the point where it can actually extend range and endurance. Also, the carriage of external tanks reduces the weight of external ordnance available for the mission.

The new Sukhoi fighter was huge. It could hardly have been otherwise given the triple requirements of a large radar antenna, long

range, and a full bag of sizeable medium-range AAMs.

Sukhoi set out to produce an optimum configuration, starting from the premise that, in any case, flight refueling was not a normal part of Soviet operations. The solution was to provide a simply enormous internal fuel capacity. There was a price to be paid. With a full fuel load, the Su-27 was g-limited, restricting maneuverability. But for tactical operations a lesser load could be used, while for extended-range missions a large amount of fuel would be used before contact was made, restoring full agility.

DEVELOPMENT

The T-10-1 flew nearly five months before the MiG-29, but its debut was flawed. Sukhoi had gained experience with FBW with the experimental high-speed research S-100 in 1972. Now they combined neutral stability with triplex analogue FBW in pitch, with mechanical backup. But problems remained. An FBW failure on May 7 1978 caused the death of test pilot Eugeny Soloviev. Then there was directional instability at speeds exceeding Mach 2, handling difficulties, and higher than predicted drag and fuel consumption. Even worse, operational studies showed the new fighter to be very much inferior to the F-15.

Sukhoi went back to the drawing board, and the result was an almost total redesign. As recalled by Vladimir Ilyushin, only the center section remained the same. Even the cockpit was made larger, although to Vladimir, built like a leprechaun, this made little difference! One of the main external differences was that the twin fins, previously mounted atop the engines and set forward as on the F/A-18 Hornet, were now located outboard on booms.

Ilyushin took the revised airplane aloft on April 20 1981, a delay of nearly four years. Its problems were not yet over; a fuel system failure caused him to eject during a test flight on September 3 1981. The leading edge slats were the cause of another disaster on December 23 when they separated in flight, killing test pilot Aleksandr Komarov. Redesigned and strengthened, they went on to perform perfectly, but problems with the FCS software, the fuel system and the radar delayed service entry to 1986, three years later than that of the MiG-29.

It was in that year that a stripped Su-27, designated P-42, flown by Viktor Pugachev and Nikolai Savodnikov, started to set a whole range of time to altitude records. As most of these had previously been set by the F-15 Streak Eagle, the West was forced to sit up and take notice.

RIGHT: The cockpit of the Su-27 is roomy and comfortable, with a raked back seat. A bubble canopy and low-cut cills give a superb view through 360 degrees.

The West's first close look at the Su-27 came in 1987, when Su-27s started to intercept Norwegian AF Orion maritime patrol aircraft. On September 13 of that year, one got a little too enthusiastic, and had one of his fins clipped by the Orion's propeller.

WESTERN DEBUT

This had an interesting sequel in 1989. A year earlier, the MiG-29 had caused a sensation with its international debut at the Farnborough Air Show, England. With little or no advance warning, two Su-27 Flankers, one of them the two-seater UB conversion trainer with taller fins, arrived at Le Bourget, Paris, for the International Air Show there in 1989. Flown by Viktor Pugachev and Eugeny Frolov, they had traveled non-stop from Moscow on internal fuel only, thus confirming their exceptional range capability.

The writer claims to be the first Western civilian ever to sit in a Flanker cockpit. HOTAS was evident, the displays were pre-1980s, but all was redeemed by the 360 degree view from the cockpit, which matched anything the West had produced. The K-36DM ejection seat was raked back at about 25 degrees, and was set high, bringing the cockpit cills well below shoulder level. The IRST sensor, set just ahead of the cockpit, appears from the ground to obstruct forward view, but in practice was hardly noticeable.

Balanced precariously on the access ladder, and accompanied by an interpreter, was Sukhoi deputy chief designer Konstantin Marbashev. After a very exhaustive presentation, I was chided for not asking whether Russian pilots enjoyed flying the type. I replied that the way that they had been "buzzing" Norwegian aircraft rendered the question unnecessary.

At close quarters, the big Sukhoi was impressive. A massive radome was flanked by huge rectangular variable intakes, raked steeply back for speeds in excess of Mach 2. Unlike the MiG-29, FOD doors were not used, but retractable titanium mesh guards protected the engines from FOD, and also from bird strike.

Le Bourget '89 also saw the first public demonstration of the "Cobra" maneuver! From level flight at moderate speed, the big fighter suddenly pitched up to an alpha of at least 100 degrees. A collective gasp went up from the crowd, who all thought that Viktor Pugachev had departed controlled flight. But no – down went the nose into level flight, and all was well.

What the Cobra demonstrated was a transitory excursion to an alpha well outside normal fighter limits while retaining full control. Western commentators eagerly seized upon it as an example of "pointability": a method of bringing weapons to bear in close combat. In fact, as Eugeny Frolov confirmed to the author several years later, it was just an aerodynamic trick, for which the alpha limiter, normally 26 degrees, had to be switched off. Only when Western reactions had been observed did the Russians start to consider the combat implications.

ABOVE: The clean lines of the basic Su-27 planform are evident from this angle. Starting with the Cobra at Le Bourget in 1989, Flankers have introduced many new airshow maneuvers.

ABOVE: Air interception missiles mounted on the folded wing of the carrier variant of Flanker. Nearest is a heat-homing R-27 Alamo variant; the others are very agile R-73 Archer dogfight missiles.

TWO-SEATERS

The Su-27UB conversion trainer has the rear seat located behind the pilot, at the cost of a small reduction in fuel capacity, and is fully combat capable. While this is a departure from normal Russian practice, it has allowed the development of a whole family of variants, although as these are generally optimized for attack missions (Su-30, Su-32, Su-34, etc.) they are outside the scope of this work.

FLANKER AT SEA

Around 1980, work began on a carrier-borne fighter, the Su-27K. Experiments with arrested landings and ski-jump take-offs began in 1984, the prototype first flew in August 1987, and the first deck landing took place in November 1989.

The navalized variant differs in detail from the standard Flanker. Safe recovery to a carrier poses unique problems, and automatic landing aids were developed. A tailhook and some structural strengthening were needed to cope with arrested deck landings, while the long tail "sting" was drastically shortened. The wings were redesigned to fold, as were the tail surfaces, and the front end sprouted canard foreplanes.

Carrier operations severely restricted take-off weight. The solution was to launch with reduced fuel, then top up in flight. Su-27Ks

As with the MiG-29, the IRST was coupled with a laser ranger and the 30mm GSh-301 cannon. No fewer than 12 hardpoints carried AAMs, giving Flanker more than adequate combat persistence. For the record, the Russians, never very good at giving military aircraft emotive names, themselves refer to the Su-27 as Flanker!

RIGHT: The Russians are noted for screwing every last ounce of capability from proven designs. The Su-35 Super Flanker differs externally from its predecessor in having canard foreplanes and a much larger radome.

were given retractable probes, the fighter refueling from "buddy" packs carried by other Su-27Ks. This was a waste of resources, but the only alternative was to carry less weaponry.

SUPER FLANKER
The first public appearance of the Su-35 Super Flanker was at Farnborough Air Show 1992, with the brochure description "triplane aerodynamic configuration fighter," which arose from the use of all-moving canard foreplanes and a conventional horizontal tail.

The Su-35 had been upgraded in all departments, not least in the search for greater agility. Relaxed stability coupled with digital FBW and canard foreplanes made it more responsive than the neutrally stable Su-27, while more powerful turbofans added to energy maneuverability. The alpha limit was increased to 30 degrees, while the extensive use of composites and aluminum lithium kept weight increases within bounds.

Super Flanker had a "glass cockpit," with color MFDs, and experiments were made with a sidestick controller, although whether this will feature on the production model is not known.

A greatly enlarged radome housed the N-011 multi-mode radar, the detection range of which was stated to be up to 86nm (160km)

against a target with an RCS of 3m². Up to 20 targets could be tracked, and six engaged simultaneously. Super Flanker also carried a rearward-looking radar with a range of 27nm (50km), which can provide guidance or a rear-firing missile, a future weaponry option. Wingtip pods housed ECM gear. Fuel capacity was increased by "wet" fins, and 14 weaponry hardpoints were installed.

SUPER-MANEUVERABILITY
At Farnborough 1996, yet another Flanker variant was on show. This was the Su-37, a fighter prototype fitted with thrust vectoring engine nozzles. This provided a measure of control in flight regimes where conventional aerodynamic is ineffective. In the hands of test pilot Eugeny Frolov, the Su-37 was put through maneuvers which were mind-blowing, notably the "Kulbit", or somersault. This started like a Cobra, but continued through a full 180 degrees until the big fighter was on its back. It then stabilized for a couple of seconds; then a quick burst of nozzle dropped the nose through. It recovered pointing in the original direction, about 30 degrees nose low, with just 32kt (59kmh) on the clock but under full control. Just an air show trick, or a significant advance in pointability? We can only wait to see if it actually enters service.

ABOVE: Flankers have astonished Western observers with demonstrations of both their range and supermaneuverability. But the question remains as to whether these are merely spectacular air show gimmicks or reliable air combat maneuvers.

Sweden

IG JAS 39 GRIPEN

A tradition of neutrality, coupled with sturdy independence, has seen Sweden produce a series of successful jet fighters. Of course, not everything is indigenous; engines, avionics and weapons have had to be imported, or license-built to foreign designs. But for a nation with a limited defense budget imposed by a population of barely 8.5 million, this is a remarkable achievement. The latest Swedish fighter is the JAS 39 Gripen.

BACKGROUND

Aware from the early 1950s that airfields were vulnerable high priority targets, Sweden adopted the concept of dispersed basing. Hundreds of short sections of roads throughout the country were selected as ad hoc airstrips, with well camouflaged dispersal areas under the trees on each side. Naturally this caused logistics problems which put premiums on reliability and maintainability in the field. More importantly, it demanded fighters with superior short-field performance. The requirement was the ability to operate from road sections only 2,625ft (800m) long.

Take-off was relatively easy; landing was the real problem. This demanded slow approach speed, good braking, and precise touchdown points. The latter could be achieved only by using a brutal high-sink rate, carrier-type, no-flare landing.

Gripen's predecessor was the Viggen, a tactical fighter built in different variants to fill the air superiority, attack, reconnaissance, and training roles. A product of early 1960s technology and operational requirements, by 1980 the Viggen was already beginning to look dated. In that year, the decision was taken to develop its successor.

The requirements were stringent. A single type of the new machine had to be able to replace Viggen in all its roles while surpassing it in areas such as air combat. It also had to be affordable in sufficient numbers, but during the 1970s costs had escalated out of sight. As there is a direct link between weight and cost, the solution had to be to produce a much smaller fighter than Viggen, but with equal or greater capabilities.

Studies showed that advances in technology – notably relaxed stability and FBW, composite construction, efficient turbofans with far

ABOVE: The JAS 39B Gripen is a two-seater. As seen here it carries six Sidewinders, but has no internal gun. As the second seat displaces fuel, endurance is less than that of the single-seater.

JAS 39A GRIPEN

DIMENSIONS: Span 27ft 7in (8.40m); Length 46ft 3in (14.10m); Height 14ft 9in (4.50m); Wing Area c275sq ft (25.54m²)

ASPECT RATIO: 2.76

WEIGHTS: Empty c12,346lb (5,600kg); Take-off c19,224lb (8,720kg)

FUEL: Internal 5,000lb (2,268kg); Fraction 0.26

POWER: 1 x Volvo/GE RM12 augmented turbofan; Maximum Thrust 18,105lb 9 (8,212kg); Military Thrust 12,141lb (5,507kg)

LOADINGS: Thrust to weight ratio 0.94; Wing 70lb/sq ft (341kg/m²)

PERFORMANCE: V_{max} high Mach 1.8; V_{max} low supersonic; V_{min} c110kt (204kmh); Operational Ceiling 50,000ft (15,239m); Climb Rate c50,000ft/min (254m/sec)

WEAPONRY: 1x27mm Mauser BK 27 cannon; 2xSkyflash or AIM-120 Amraam; 4xAIM-9 Sidewinder.

USERS: Sweden, possibly Hungary.

ABOVE: Gripen launches a medium range AAM during weapons trials. AIM-120 Amraam is the preferred weapon for the future, backed by Sidewinder for close combat.

RIGHT: The leader peels away during low-level training over the Swedish countryside. Note that the canard foreplanes are set behind the pilot's eyeline.

BELOW: Armorers winch missiles onto an underwing pylon before a sortie. Gripen is designed for rapid turnaround with much of the work done by conscripts.

higher thrust/weight ratios, and advanced avionics - would allow a payload/range performance comparable to that of the Viggen, for half the take-off weight of the latter. In the air combat arena, maneuverability would be far superior. Small size had a spin-off, in that RCS would be minimized without positive stealth measures being incorporated.

DEVELOPMENT

In 1980 *Industri Gruppen* JAS was formed, a consortium of Saab-Scania, Volvo Flygmotor, Ericsson, and FFV Aerotech, to develop what was to become the JAS (Jakt, Attack, Spaning) 39. Basically it was to be an air superiority fighter (*Jakt*), with *Attack* and *Spaning* (Reconnaissance) as secondary missions.

With a target weight of eight tons settled for the new fighter, a suitable engine was the first priority. The choice fell on General Electric's F404, selected for the F/A-18 Hornet. GE and Volvo Flygmotor worked to screw an extra 2,000lb (907kg) out of it. Mass flow was increased by five percent, and turbine entry temperature raised by 105 degrees Celsius, while resistance to birdstrike was improved. But problems had to be overcome on the way. What was now the RM12 suffered from slow thrust build-up on take-off, cracking compressor blades, and afterburner problems.

With the engine selected, the next task was to wrap the new fighter around it. The first consideration was the intake. A chin location had many advantages, but posed design problems with such a small fighter. The alternative was fixed-geometry side inlets. While sideslip would cause distortion of the airflow, the RM12 was known to be tolerant. A further advantage was reduced RCS from head-on.

A canard delta configuration was adopted for maximum agility consonant with high acceleration and performance; as with France's Rafale, the all-moving canards were close-coupled, and located at the top of the inlets, aft of the pilot.

The evolved Saab 39 has a mid-position wing, almost entirely of composite construction, which allows adequate stores clearance. Wing/body blending is minimal, and root extensions absent. Two-piece slats occupy most of the leading edge, separated by a dogtooth discontinuity against spanwise flow, while two-piece elevons make up the entire trailing edge. The tips are cropped to allow carriage of AAMs.

About 30 percent of the structure is made up of composites, including the vertical tail. The canopy is faired into a dorsal spine, which carries control runs. The main gears, stressed for hard landings, retract into the fuselage aft of the intakes. Unlike Viggen, Gripen has neither a braking parachute nor thrust reversers. Its short landing run is entirely dependent on a combination of efficient braking coupled with aerodynamic downloads, speed brakes on each side of the aft fuselage, and deflecting the canards steeply downwards to create drag.

The cockpit is modern, with HOTAS, three MFDs, and a wide-angle diffraction-optics HUD fronted by a wrap-around windshield. The ejection seat is raked at a 22 degree angle. Ericsson make both the radar and the self-protection suite. The PS-05/A pulse-Doppler multi-mode radar was developed specifically for Gripen. Few details have been released, but it is known to have at least five main air-to-air modes, plus a radar-gun aiming mode. It provides monopulse SARH for Skyflash AAMs, and midcourse guidance updating for Amraam and MICA. The planar array scanner is small, just 23.62in (600mm) diameter, but given a 1kW power output, reinforced by low noise preamplifiers, a search range of about 54nm (100km) is a reasonable guess.

FLIGHT TESTING

The maiden flight of the first Gripen prototype took place on December 9 1988, with chief test pilot Stig Holmström at the controls. Much of the preceding two years had been spent in validating the software for the triplex digital FBW system, even though this had been flown in a Viggen testbed since 1982. Stig himself had spent about 1,000 hours in the simulator, and on landing he commented that, while the aircraft flew well, the control system was a bit too sensitive.

That all was not well was amply demonstrated on February 2 1989. As test pilot Lars Rådeström came in to land, the Gripen started to pitch, hit the runway hard, and came to rest inverted. It was a write-off, although fortunately Lars escaped with relatively minor injuries.

Software control laws were found to be at fault, and amendments were made. When in May the flight test program was resumed, Gripen was much more stable. In September 1992, Gripen made its international debut at Farnborough Air Show, England.

Just when all seemed to be going well, Lars Rådeström again drew the short straw.

During a public display over Stockholm on August 8 1993, Gripen became uncontrollable during a shallow turn, and Lars ejected. Fortunately no-one on the ground was injured by the crashing aircraft. This major setback was not righted until 1995, when the new P11 flight control software was passed as ready.

TWO-SEATER

First flown on April 29 1996, the two seater JAS 39B is a combat trainer. To make space for the second cockpit, the fuselage is 26in (655mm) longer, fuel has been reduced, and the gun deleted to save weight. The first JAS 39B entered service in April 1998.

THE FUTURE

The first Gripen unit was F7 Wing at Såtenäs, the second squadron of which was declared operational late in 1997, and the Gripen build-up will continue for many years.

In an era where a thrust/weight ratio exceeding unity is almost obligatory for a fighter, Gripen is currently somewhat marginal. More powerful engines are under consideration for the JAS 39C, and IG is also currently working on a thrust-vectoring demonstration program with Germany and the US Navy. Finally, flight refueling, not a Swedish requirement, is to be offered for export models.

ABOVE: Four medium range AAMs, either Amraam or Skyflash, with two wingtip-mounted Sidewinders, give Gripen a credible air-to-air capability.

Taiwan

AIDC A-1 CHING KUO

The entry of Taiwan into the fighter business stemmed from the claims and counterclaims of the People's Republic of China and the Republic of China (Taiwan) to each other's territory. Initially regarded as a bulwark against communism, Taiwan was lavishly equipped with American combat aircraft, but was sidelined when the Cold War thawed. New US combat aircraft were no longer available, a factor which helped sink the Northrop F-20, but ambiguities in US restrictions meant that technical assistance remained a possibility.

RIGHT: Three two-seater Ching Kuo conversion trainers seen above the clouds, carrying tip-mounted Sky Sword AAMs. Sky Sword I owes much to Sidewinder which it closely resembles.

Faced with being saddled with obsolescent tactical aircraft for the foreseeable future, Taiwanese authorities made the decision in 1992 to develop an indigenous fighter with American technical assistance, and the design was finalized in 1985. The resulting fighter duly emerged as the Ching Kuo.

With a limited choice of powerplants, the Garrett (now ITEC) TFE 1042 augmented turbofan was selected, but as the thrust of this was extremely limited a twin-engined configuration was necessary. General Dynamics (now Lockheed Martin) provided technical assistance with the airframe.

Fixed-geometry side intakes, with a passing resemblance to those of Rafale, were located beneath extended strakes. The lack of a chin intake gave a shallow fuselage depth. This, allied to the Lockheed Martin influence, gave Ching Kuo the side-on appearance of an anorexic F-16. In other areas the similarity was pronounced; differences were a downwards-raked fin top, forward-swept wing trailing edges, giving a comparable area for a shorter span, and the absence of ventral fins. Given the shallow depth of fuselage, the wing is almost shoulder-mounted, with a high degree of wing/body blending.

The single-piece cockpit transparency of the Fort Worth fighter was not duplicated. Ching Kuo has a conventional wrap-around windshield and canopy bow. A cockpit mock-up was shown at the Paris Air Show in 1994. This was almost identical to that of the American fighter, with two MFDs, a steeply raked ejection seat, and a sidestick controller. It is rather smaller, however, since it is sized for Taiwanese pilots rather than Westerners, as AIDC senior test pilot Goang-Hwa Jong (otherwise known as Robert) pointed out to the writer. This is a natural restriction on its export potential.

The flight control system is triplex digital FBW, with many similarities to that used in the Gripen. The crash of the latter in February 1989 delayed the maiden flight of Ching Kuo until May 28 1989, when it was piloted by Wu Kang Ming. That all was not well with the FCS was demonstrated by a landing accident similar to that of Gripen on October 29 1989. The second pre-production machine was lost on June 12 1991, due to tail failure caused by severe vibration.

Like all modern fighters, Ching Kuo has a two-seater conversion trainer, based on the minimal change principle. The second seat displaces the leading fuselage fuel tank; the canopy is extended; but there is no increase in fin area to compensate for the extra keel area.

Radar is the indigenous Golden Dragon 53, based on the Westinghouse APG-67(V) multi-mode pulse-Doppler, with a maximum search range of at least 30nm (56km) in look-up, and 20nm (37km) in look-down mode. AAMs carried are the Sky Sword I, which is based on Sidewinder, and Sky Sword II, a SARH missile carried semi-recessed beneath the fuselage.

IN SERVICE

To begin with 250 Ching Kuos were ordered, and the fighter achieved initial operational capability in January 1995. But the loss of two aircraft in mysterious circumstances caused the fleet to be grounded, pending modifications to the fuel system. But even before this, US restrictions were lifted, and F-16A/Bs became available. In addition, Taiwan has ordered 60 Mirage 2000-5s from Dassault. Only 60 Ching Kuos have been delivered, and production has now ceased.

A-1 CHING KUO

DIMENSIONS: Span 30ft 10¼in (9.42m);
Length 46ft 7¾in (14.20m); Height 15ft 6in (4.72m);
Wing Area 260sq ft (24.20m²)

ASPECT RATIO: 3.67

WEIGHTS: Empty 14,300lb (6,486kg);
Take-off 21,000lb (9,526kg)

FUEL: Internal 4,650lb (2,109kg); Fraction 0.22

POWER: 2xITEC TFE1042-70 augmented turbofans;
Maximum Thrust 9,500lb (4,309kg);
Military Thrust 6,060lb (2,749kg)

LOADINGS: Thrust to weight ratio 0.90;
Wing 81lb/sq ft (394kg/m²)

PERFORMANCE: V_{max} high Mach 1.65; V_{max} low Mach 1.04;
V_{min} n/a; Operational Ceiling c55,000ft (16,763m);
Climb Rate c50,000ft/min (254m/sec)

WEAPONRY: 1x20mm M61 cannon, 4 x Sky Sword I IR
homers, 2xSky Sword II SARH homers

ABOVE: The influence of Fort
Worth on the design of Ching Kuo
is obvious from this angle, but
lack of a suitable powerplant
forced a twin-engined layout to
be adopted.

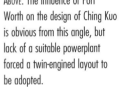

USA

NORTHROP GRUMMAN F-14D TOMCAT

By the 1960s, the greatest threats to the carrier groups of the US Navy were fast jet bombers armed with long range stand-off anti-ship missiles (ASMs). The best defense against these was to intercept the bombers before they could launch their weapons. The fleet air defense fighter of the time, the ubiquitous F-4 Phantom, was simply not good enough. What was needed was a fighter that could remain on station at least 150nm (278km) out from the carrier for extended periods, and destroy multiple targets from far beyond visual range. It needed supersonic dash speed both to run down fleeing opponents when necessary, and to reinforce existing patrols from the deck. It needed agility for other missions such as air superiority and escort. Finally, to be carrier compatible, it needed a relatively slow approach speed at a shallow alpha.

The Grumman Aircraft Corporation had been commissioned to navalize the General Dynamics F-111 attack aircraft, but it soon became obvious that this was an overweight and unmaneuverable non-starter. Even before it was canceled, Grumman had started preliminary design studies for an alternative. By 1970, this had emerged as the F-14 Tomcat.

DEVELOPMENT

Long range AAMs and a compatible radar and weapons control system developed for the defunct F-111B were specified for the F-14. These needed a big fighter to carry them, making two powerful engines essential. The only real candidate at that time was the Pratt & Whitney TF30 turbofan, as used in the F-111, but it was known to be sensitive to disturbed airflow. To minimize this, a "straight-through" configuration was used, with a huge radome and two-man cockpit in a frontal nacelle, flanked by variable intakes needed for Mach 2 plus. The engines, following straight lines from the intakes, were thus well spaced, and were joined by a "pancake" aft fuselage. This had the added advantage of giving extra lifting area.

This was just as well. The combination of supersonic speed, extended loiter time, and a low speed, low alpha approach, was met by a variable-sweep wing. But these are inevitably small, giving a high wing loading; the additional lifting area provided by the pancake offset this by a remarkable amount.

The wing design was a tour de force. Sweep angles were infinitely variable, computer controlled for optimum settings in all flight

ABOVE: The flat landing angle of the Tomcat, so essential for carrier operations, is clearly apparent as this F-14 of VF-211 Checkmates comes aboard USS *Nimitz* in May 1988.

F-14D TOMCAT

DIMENSIONS: Span (max/min) 64ft 1½in/38ft 2½in (19.55m/11.65m); Length 61ft 11in (18.87m); Height 16ft 0in (4.88m); Wing Area 565sq ft (52.50m²)

ASPECT RATIO: (max/min) 7.28/2.58

WEIGHTS: Empty 41,780lb (18,951kg); Take-off 61,200lb (27,760kg)

FUEL: Internal 16,200lb (7,350kg); Fraction 0.26

POWER: 2xF110-GE-400 augmented turbofans; Maximum Thrust 27,080lb (12,283kg); Military Thrust 16,610lb (7,534kg)

LOADINGS: Thrust to weight ratio 0.88lb; Wing 108lb/sq ft (529kg/m²)

PERFORMANCE: V_{max} high Mach 1.88; V_{max} low Mach 1.20; V_{min} c100kt (185kmh); Operational Ceiling 53,000ft (16,154m); Climb Rate c48,000ft/min (244m/sec)

WEAPONS: 1x20mm M61 cannon with 675 rounds; 4xAIM-120 Amraam and 4xAIM-9 Sidewinder;

(F-14A, as artwork, AIM-54 Phoenix, AIM-7 Sparrow, AIM-9 Sidewinder)

USERS: Iran, USA.

regimes. To minimize deck parking space, oversweep was used, rather than traditional wing folding. Leading edge slats and trailing edge flaps provided extra lift for maneuver and landing. At supersonic speeds, vanes in the wing gloves extended to compensate the tail download.

Differentially moving tailerons provided pitch and roll control, the latter augmented by wing spoilers at subsonic speeds, while twin fins and rudders were supplemented by shallow ventral strakes.

WEAPONS SYSTEM

The weapons system was the trend-setting Hughes AWG-9, with six large AIM-54 Phoenix AAMs giving an unprecedented long-range kill capability. For the air superiority mission, the normal load was four AIM-7 Sparrow and four AIM-9 Sidewinder AAMs.

The AWG-9 pulse-Doppler radar could simultaneously track up to 24 targets, and provide targeting information for six Phoenix launched in quick succession. Detection range was exceptionally long, nominally 115nm (212km) against fighters, and more than 65nm (120km) against cruise missiles. It sported a variety of air-to-air modes: pulse-

Doppler Search, Range While Search, and Track While Scan, plus close combat modes.

TOMCAT ALOFT

The F-14 first flew on December 21 1970, but this prototype was lost on its second sortie nine days later, due to a double hydraulic failure. Testing resumed on May 24 1971 with the second prototype.

Carrier landing trials took place in 1972. Deck landings were tricky, due to sluggishness in roll at low speeds, but in the air the Tomcat handled well although in extreme circumstances it could enter an irrecoverable flat spin. This the US Navy had to live with.

INTO SERVICE

The Tomcat entered service with Replacement Air Group VF-124 in 1973. Although a two-seater, the F-14 had no provision for flight controls in the rear cockpit, although this does not seem to have caused problems. The first two operational Tomcat squadrons, VF-1 Wolfpack and VF-2 Bounty Hunters, embarked on USS *Enterprise* late in 1974.

Meanwhile the AWG-9/Phoenix combination was put to the test. April 1973 saw it used against a bomber-sized target flying at Mach

BELOW: The use of afterburning for takeoff reveals that this is an F-14A. Aircraft engined with the more powerful GE F110s can deck-launch using military thrust only.

1.5. Tracking began at 132nm (245km); a Phoenix was launched at 110nm (204km). Using a high trajectory to increase range, it passed within lethal distance of the target, having traveled 721/2nm (134km).

Multiple-target engagement trials were held, culminating on November 21 1973 when a Tomcat of the Joint Evaluation Team intercepted a simulated raid of six drones flying at altitudes of 22-24,000ft (6,705-7,315m) and at varying speeds. From an altitude of 28,400ft (8,656m), the Tomcat detected the intruders at more than 85nm (158km), assigned each missile to a target, and from a range of 31nm (57km) ripple-fired all six AIM-54s within 38 seconds. The result was four direct hits, one miss, and one no-test. In all, 155 AIM-54s were launched, with a success rate of 92 percent.

ENGINE PROBLEMS

The TF30-powered F-14A was initially to be an interim machine, pending availability of the more powerful and robust F401 engine, also from Pratt & Whitney, which was to power the definitive F-14B. This was killed off by astronomic cost escalation, which threatened the entire program.

During the test and evaluation phase, the F-14A had been flown by pilots of above average ability, in carefully graduated trials. When it reached the squadrons it was a totally different matter. In simulated combat, the squadron pilots were continually working the throttles. The TF30 could not take such hard usage; compressor stalls were frequent, compressor blades were shed, and all too many aircraft were lost to engine-related causes. Tomcat pilots had to learn to fly the engine as well as the airframe to avoid the worst effects of this.

F-14B AND F-14D SUPER TOMCAT

This situation could not be allowed to continue. By 1984, the F-14 was a prime candidate for upgrading, the first step of which was to fit General Electric F110 turbofans. These greatly increased the thrust/weight ratio, giving tremendous performance advantages and range increments. This entered service in November 1989 as the F-14+, but as this designation was not compatible with US Navy computer programs, it was later amended to F-14B.

The next step was an avionics upgrade. A new radar, the Hughes APG-71 used digital processing to produce a six-fold increase in capability. Low Light Television, already standard on the F-14A, was retained, but was now supplemented by an IRST, which could be slaved to the radar. Many other technical

innovations were included, and MFDs replaced many conventional instruments. The circular Tactical Information Display in the rear seat was retained; it can now display out to 400nm (741km), which allows it to accept long range contacts provided from other sources via data link.

Following the dissolution of the Soviet Union, Tomcat production ceased in July 1992. Only 37 new-build F-14Ds, plus 18 remanufactured aircraft reached the Navy, with a similar number of F-14Bs.

Tomcats have seen little air combat. In August 1981 they shot down two Libyan Su-22s; in January 1989 it was the turn of two Libyan MiG-23s. In the Gulf War of 1991, lack of opportunity stopped them adding to their laurels.

ABOVE: Two Tomcats of VF-143 Pukin' Dogs line up for a deck landing on the carrier USS *Eisenhower*. Note that arrestor hooks are already down, even though the landing gear has not yet been lowered.

BELOW: An F-14 of VF-84 Jolly Rogers is "fed to the cat" on USS *Theodore Roosevelt*, prior to launching. On the far side can just be seen a F/A-18 Hornet, a much smaller aircraft.

USA
BOEING F-15C EAGLE

If at any point during the Cold War, hostilities had erupted between East and West, one thing was certain: NATO forces would have been heavily outnumbered. While this was worrying, American fighters had what was then assessed as a 15-year technological lead over the Soviet Union. Quality would offset quantity. Or would it?

In Vietnam, the world's most advanced fighter at the time, the F-4 Phantom, was opposed by cheap and cheerful MiG-21s, and by even more basic MiG-17s. In theory the Phantom, with its eight AAMs and BVR capability, should have carved swathes through the North Vietnamese MiGs, but in practice this did not happen. At times the kill/loss ratio was actually adverse from the United States' point of view.

ABOVE: F-15s of the 21st Fighter Wing on patrol high over Alaska. Conformal tanks show these to be an F-15C and, nearest the camera, a two-seater F-15D. Two-seaters are fully combat capable.

The situation was exacerbated by faulty intelligence. The Soviet Union's MiG-25 Foxbat was known to have outstanding speed/altitude performance; it was erroneously assumed that it was also a maneuverable air superiority fighter. Furthermore, it was also assumed to have been ordered in large numbers; confusion in this case arose from orders for the MiG-23, a lesser bird.

Faced with what appeared to be a quantum leap in Soviet fighter development, the USAF demanded a counter. This had to be a sophisticated weapons system, which could combine outstanding performance with

F-15C EAGLE

DIMENSIONS: Span 42ft 9½in (13.05m);
Length 63ft 9in (19.43m); Height 18ft 5½in (5.63m);
Wing Area 608sq ft (56.50m²)

ASPECT RATIO: 3.01

WEIGHTS: Empty 28,600lb (12,973kg);
Takeoff 44,500lb (20,185kg)

FUEL: Internal 13,455lb (6,103kg); Fraction 0.30

POWER: 2xF100-P-220 augmented turbofans;
Maximum Thrust 23,770lb (10,782kg);
Military Thrust 14,590lb (6,618kg)

LOADINGS: Thrust to weight ratio 1.07
Wing 73lb/sq ft (357kg/m²);

PERFORMANCE: V_{max} high Mach 2.50; V_{max} low Mach 1.20;
V_{min} 100kt (185kmh); Operational Ceiling 65,000ft
(19,811m); Climb Rate 50,000ft/min (254m/sec)

WEAPONRY: 1x20mm M61A
cannon with 675 rounds
8 AAMs, AIM-120 Amraam
and AIM-9 Sidewinder

USERS: Israel, Japan,
USA, Saudi Arabia.

outstanding agility. It was to be a single-seater with Mach 3 capability, although as the demand for agility conflicted with the Mach 3 requirement, a compromise figure of Mach 2.5 was agreed.

Weapons load was to be the same as that of the Phantom: four AIM-7 Sparrows and four AIM-9 Sidewinders. Excluding launch rails, this load weighed about 2,750lb (1,247kg), which made two engines essential to achieve a satisfactory thrust/weight ratio. Given this, a large wing area (to keep wing loading, and therefore turn rates, to acceptable levels, combined with moderate aspect ratio to give reasonably fast roll rates) was needed. The result was inevitably a large and heavy aircraft.

DEVELOPMENT

The contract for what was to become the F-15 was awarded to McDonnell Douglas (later

part of Boeing) in December 1969. Competition between Pratt & Whitney and General Electric resulted in the selection of the former's F100 twin powerplants in 1970. Large turbofans, they were developed specifically for the F-15.

The F100 was far more advanced than the TF30 of the F-14 Tomcat; a more efficient compressor and 27 percent higher turbine entry temperatures gave much more thrust in both military and augmented power, even though the F100 was smaller and lighter.

The requirement for Mach 2.5 gave rise to sharply raked two-dimensional variable inlets with movable ramps controled by the air data computer. Unique pivoting hoods (4deg up, 11deg down) were used to keep the inlets pointing into the local airflow during maneuvering; like canard foreplanes, these also created a download at supersonic speeds to unload the tail surfaces.

Below: The Israeli Air Force has operated F-15s for more than 20 years with great success. Since 1979, more than 40 Arab victims have fallen to Israeli pilots.

The shoulder-mounted wing was virtually a cropped delta, set at zero incidence to maximize level flight acceleration. The leading edge was fixed; the trailing edge had ailerons outboard and flaps inboard. Inboard, the three main spars were built from titanium; battle damage to one would allow a safe return.

Like the slightly earlier Tomcat, the Eagle design had twin fins and rudders, and differentially moving stabilizers mounted on booms outside the line of the engines. These had a notched leading edge, a remedy for flutter.

Radar was the Hughes APG-63 multi-mode pulse-Doppler, optimized for the air-to-air regime. Typical detection range against fighter-sized targets was something in excess of 86nm (159km). While this lacked the range of the F-14's AWG-9, and its multiple target attack capability, it was greatly superior in the field of low closure-rate target detection and tracking. Attack mode was basically look-shoot, look-shoot, "stepping through" targets on demand or as it went.

Cockpit design was trend-setting. A large clear-vision canopy gave excellent 360 degree coverage, unlike its Phantom predecessor. But, whereas with the Phantom and its close contemporary the Tomcat the workload involved in flying the airplane and operating the avionics and weapons system was too much for one man to cope with, the Eagle was a single-seater by design. A new approach was needed.

The solution was HOTAS - Hands On Throttle and Stick - in which all controls which the pilot would need in combat were situated under his hands on one of the two major flight controls. The twin throttle levers contained switches which controlled the microphone; IFF interrogation; target designator; gunsight reticule; radar elevation; weapon selection; chaff and flare selection; weapon selection; and speed brake. On the control column were mounted the gun trigger/HID camera; pickle (i.e., weapons launch); radar auto-acquisition; missile seeker head cage/uncage; and trim.

No longer did the pilot have to fumble in the cockpit for the correct switch, nor risk taking his eyes off a fleeting target at the limits of visibility or be distracted from the tiny radar screen in the top left-hand corner of the instrument panel; everything he needed was now immediately to hand. This was not done without cost, however; it needed the manual dexterity of a piccolo player to get the best out of the system. It called for training, training, and still more training.

When the F-15 first entered service in Europe, the author spoke to USAF Aggressor pilot Bill Jenkins who assured him that it was possible to sneak up and fly wing while an Eagle pilot was hypnotized by all the cockpit magic. But fortunately this state of affairs was temporary. One thing that must be said: HOTAS was so successful that every other important fighter-building nation - Britain, France, Sweden, Israel, Russia, etc. - followed suit.

First flight and into service
The first flight of the F-15 was ignominious, in the hold of a C-5 Galaxy from McDonnell Douglas's St. Louis plant to Edwards AFB in California. It took to the skies under its own power in the hands of chief test pilot Irving Burrows on July 27 1972. A few problem areas were identified and fixes found, but in the main these were not serious. Before long, the USAF recognized that it had a winner, and in 1975 a stripped version, the Streak Eagle, set a hatful of time-to-altitude records, many of which had been previously held by the Russian MiG-25 Foxbat.

Every seventh F-15 was built as a two-seater. Originally designated TF-15, this has since become the F-15B. Usually this is done at the expense of fuel, but this was not the case with the F-15B, which is fully combat-capable.

The Eagle entered service in November 1974, and achieved initial operational capability with 1st TFW at Langley in 1976. Generally regarded as easy to fly, if a little sensitive in pitch, the Eagle lacked only a little in rate of roll; otherwise it was a fighter pilot's dream. Just one problem obtruded: in simulated combat the F100 was insufficiently durable to

BELOW: The cockpit of an F-15A, with dozens of dial instruments and a tiny radar display at top left. On the other hand, the many switches and buttons on the control column are part of the HOTAS system.

cope with repeated throttle slams from one extreme to the other; stagnation stalls and turbine failures were common. The author recalls seeing a turbine failure during the pre-take-off run-up when flames came out of every conceivable orifice!

Nor did augmentation always work as advertised; many pilots left it on minimum setting rather than have it fail to light. The effect of this was that fuel was used at higher than predicted rates, making the F-15A rather more short-legged than it should have been.

IMPROVING THE BREED

Even as the F-15A was undergoing its initial flight trials, its designers were at work on an improved model. Extra fuel tankage was installed in the forward fuselage and the wings, increasing basic capacity by 1,820lb (826kg). But the most innovatory approach to the fuel problem was to install conformal fuel tanks on the fuselage sides beneath the wings. These, which can be fitted or removed in just 15 minutes, hold a massive 9,750lb (4,423kg) of fuel, for only a marginal drag increase. They can also be fitted with IR sensors, ECM kit, etc.

Most other improvements concerned the APG-63 radar. A programmable signal processor controlled the modes via software rather than circuitry, and greater computer capacity allowed new modes to be added. Among these were Raid Assessment, which could pick out individual aircraft from a close formation at up to 40nm (74km), and Track While Scan. Meanwhile, plans were afoot to treble the speed of data processing to 1.4 million operations per second.

The original slogan of the design team at St. Louis had been, "Not a pound for air to ground," which resulted in an outstanding air superiority fighter. But, in the time-honored manner of all air forces, it was not long before the Eagle was cleared to carry air to ground ordnance, and a beefed-up landing gear became necessary to increase the maximum allowable take-off weight.

The result was the F-15C, which replaced the A on the production line in 1978, and made its first flight on February 27 1979, followed four months later by the two-seat F-15D. Without the conformal tanks, the only external difference between the A and C and the B and D was the landing gear. The speed of events had overtaken radar development, and much of this, and other avionics improvements, were carried out retrospectively as part of a Multi-Staged Improvement Program starting in 1983.

The much more durable -220 engine entered service with the Eagle from 1986, giving greater reliability at the expense of slightly reduced thrust. Given the weight growth of the Eagle, this was a retrograde step; even so by any yardstick performance and capability was still high enough to outmatch any adversary aircraft then in service. But gradually a few tweaks restored the balance. What the fighter really needs is the far more powerful -229 engine, which powers the F-15E, a dedicated two-seater interdiction airplane which falls outside the scope of this book.

THE EAGLE IN PEACE AND WAR

Stressed for +9g and -3g, the F-15 was maneuverable in its day, but sadly that day is now passing. Roll rate is sluggish, while alpha is limited to 20 degrees. Sustained turn rate at 15,000ft (4,572m) is just 11.8deg/sec at Mach

0.9, while transient turn rate at the same combination is 16.5deg/sec.

In close combat the Eagle's sheer size becomes a disadvantage. It can make a fast undetected approach, but the instant it turns and presents its planform it becomes visible from miles away. Maneuvering an Eagle has been called "tennis courting" for this reason!

Where the Eagle really scores is in performance. Its high thrust/weight ratio can accelerate it by up to 100kt (185kmh) in three seconds, and that is in level flight, without unloading. In close combat, it can disengage vertically from the melée, an attribute which for many years made it unique in the fighter community.

Another strong point is its weapons system, which has undergone almost continuous upgrading. Some 350 F-15C/Ds are scheduled to be retrofitted with a new and very advanced radar from the turn of the century. Optimum air-to-air weapons load is now eight AIM-120 Amraam, or four Amraam and four Sidewinders.

The F-15 has seen action with three air forces. Its first aerial victories were scored on June 27 1979, when four Syrian MiG-21s were downed by Israeli Eagles. More victories followed, culminating in the Beka'a action of 1982.

In June 1984, it was the Saudis' turn, when F-15s knocked down two Iranian F-4 Phantoms over the Gulf.

The Gulf War of 1991 saw F-15s score the vast majority of the air-to-air kills in that conflict, many of them MiG-29s. At the time of writing they have also accounted for four Serbian MiG-29s in the Balkans. No F-15s have been lost in air combat.

ABOVE: "Nodding" intakes down, indicating low speed and high alpha, an F-15C of the 21st Fighter Wing takes on fuel from a KC-10 tanker.

USA

LOCKHEED MARTIN F-16 FIGHTING FALCON

ABOVE: Conceived as a close combat fighter, the F-16 has since been developed as a multi-role aircraft. These Block 50 Vipers, bearing the tail code of the 8th Fighter Wing, based at Misawa in Japan, are armed for the defense suppression mission.

The Fighting Falcon may be the official name of the F-16, but it is far more widely known as the Viper, or Electric Jet! Its origins date back to the time when the quality versus quantity debate was at its height. The "worst case" scenario was war in Central Europe between NATO and the Warsaw Pact, in which the superb F-15 Eagle would be heavily outnumbered by hordes of agile Russian-built fighters. The fear was that the superior American fighters would be overwhelmed.

In the Pentagon, a small faction known as the Fighter Mafia addressed the problem. They concluded that the solution was the hi-lo mix: a nucleus of very capable fighters, backed by a horde of austere, affordable dogfighters which would give something approaching numerical parity.

F-16C FIGHTING FALCON

DIMENSIONS: Span 31ft 0in/ 9.45m;
Length 49ft 3½in/15.02m; Height 16ft 8½in/ 5.09m;
Wing Area 300sq.ft/27.87m²

ASPECT RATIO: 3.20

WEIGHTS: Empty 18,600lb/ 8,437kg;
Takeoff c27,500lb/12,474kg

FUEL: Internal 7,162lb/ 3,249kg; Fraction 0.26

POWER: 1xF100-P-129 augmented turbofan;
Maximum Thrust 29,100lb/13,200kg;
Military Thrust 17,800lb/ 8,074kg

LOADINGS: Thrust 1.06; Wing 92lb/sq.ft/448kg/m².

PERFORMANCE: V_{max} high Mach 2.00 plus; V_{max} low
Mach 1.20; V_{min} n/a; Operational Ceiling 50,000ft/
15,239m plus; Climb Rate 50,000ft/min-254m/sec.

WEAPONRY: 1x20mm M61A cannon with 511 rounds;
4xAIM-120 Amraam and 2xSidewinders or
6xSidewinders or Python

USERS: Bahrain, Belgium, Denmark, Egypt, Greece,
Holland, Indonesia, Israel, Norway, Pakistan, Portugal,
Singapore, South Korea, Taiwan, Turkey, United Arab
Emirates, USA, Venezuela.

The result was the Light Weight Fighter (LWF) program, a technology demonstration to explore possibilities. Contracts for this were placed in 1972; one with General Dynamics (as it then was) now Lockheed Martin, the other with Northrop.

Requirements were few: a load factor of 7.33g while carrying 80 percent fuel, and good turning performance between Mach 0.9 and Mach 1.2. Simplicity and agility were the keynotes; armament was envisaged as two Sidewinders and a 20mm M61A Vulcan cannon, backed by a simple ranging radar.

GD set up a secure "Skunk Works" type operation to develop the LWF. The required level of agility was to be provided by a combination of a very high thrust/weight ratio by wrapping the smallest possible airframe around the largest available engine, and a modest wing loading.

The engine selected was the Pratt & Whitney F100, then under development for the F-15. With no requirement for Mach 2, a plain chin-type inlet was selected, with the forebody acting as a compression wedge at high alpha. This avoided disturbed airflow which would have resulted from conventional side intakes.

GD increased the loading to 9g with full internal fuel. They also recognized that a simple ranging radar was inadequate, and made provision for a small multi-mode type. In a trend-setting move, trim drag at high Mach numbers was overcome by locating the center of gravity forward of the center of lift. Under normal circumstances, this would have made the airplane uncontrollable, but computerized stability augmentation via a quadruplex analog fly-by-wire (FBW) system was used to control the sensitive handling that resulted. This also gave significant weight savings by allowing the elimination of hydro-mechanical control systems. To give carefree handling, alpha and g-limiters were built into the flight control computers.

The wing itself was of moderate sweep and used variable camber, with flaps on both leading and trailing edges to optimize lift in all flight regimes, while leading edge root extensions were used to form vortices across the upper wing surfaces to improve lift at high alpha. The planform was selected to give an optimum compromise between wing loading and aspect ratio.

The mid-set wing was blended into the fuselage. This added to structural strength, provided extra volume, and fortuitously reduced RCS. It was extended aft, where it provided shelves for the differentially moving stabilizers. A single fin and rudder was centrally mounted, supplemented by ventral strakes for stability at high alpha. Construction was almost entirely conventional, with little use of titanium or composites.

The cockpit was radical. To help the pilot resist gray-out at high-g forces, the ejection seat was raked back at a 30 degree angle, and the pilot's heel-line was raised, at the cost of reducing the dash space available for instruments. All-round visibility was exemplary, with cills cut low. There was no separate windshield; the thick and heavy tear-drop canopy was in one piece, and lacked a canopy bow on which to hang rear vision mirrors. The most radical change was the use of a sidestick controller. As with the F-15, switches required at critical flight points were situated on the throttle and stick.

BELOW: The planar array antenna of the APG-68 radar, used in the F-16C/D. Very obvious is the single-piece bulged canopy which gives superb all-round vision from the cockpit. The slight gold tint to the transparency is a stealth measure, designed to absorb radar emissions.

FLYOFF

The first flight of the YF-16 took place at Edwards AFB on January 20 1974. During a high-speed taxi trial, test pilot Phil Oestricher ran into trouble with oscillations due to too-high gain on the FBW, and lifted off to sort out the problem in the air. This was quickly adjusted.

In April, the LWF demonstration became the basis for a new air combat fighter (ACF) – which became the lightweight fighter (LWF). A thorough flight evaluation between the YF-16 and the Northrop YF-17, part of which involved mock combat between the two, would decide the outcome.

It was a closely run contest, but the evaluation pilots found that they were winning in the YF-16 more often than not. In most regimes the YF-16 was the better-turning and faster-accelerating of the two; but the deciding factor was transient performance. Faster pitch and roll rates gave the advantage to the YF-16.

Selection of the YF-16 as the new ACF/LWF was announced in January 1975. The decision was not hindered by the fact that the YF-16 was potentially more affordable than its twin-engined Northrop rival.

The YF-16 could have entered production "as was" but, as GD had foreseen, the USAF wanted more capability. In addition, four European nations - Belgium, Denmark, Holland, and Norway - were all very interested, and they too wanted more, specifically multi-role, capability. It was back to the drawing board!

The Westinghouse APG-66 pulse-Doppler multi-mode radar was selected for the F-16. This had four air-to-air modes, and ranges against a fighter-sized target were 39nm (72km) in Uplook and 26nm (48km) in Downlook. The nose was lengthened by 7in (178mm) and widened by 4in (102mm) to accommodate it.

Two more hardpoints were added, and as heavy external loads moved the center of gravity forward the horizontal tail area was increased by 15 percent. The fuselage was lengthened by 14in (356mm) and the fin raised 5in (127mm). All this was not done without penalty; total weight growth was nearly 2,000lb (907kg). Wing area was increased by 20sq ft (1.86m2). While this kept the wing loading down, span remained the same, with a small reduction in aspect ratio. The ability of the F-16 to switch from air superiority to ground attack caused it to be labeled the "Texan Swing Fighter."

European evaluation continued, and orders were announced at the end of the Paris Air Show in 1975, an event notable for a breathtaking display by GD test pilot Neil

LEFT: The Block 60 F-16 (F-16ES) differs from previous models in that it has conformal fuel tanks above the wing/body blending. This increases endurance at the expense of maneuvrability and performance.

Anderson in the YF-16. A new production line was put in place at Fort Worth, and work started on the initial contract for 15 full-scale development aircraft; the first of these flew from Fort Worth in December 1976.

Naturally there was a two-seater conversion trainer which, in keeping with standard Western practice, was fully combat-capable, although with 17 percent less internal fuel to make room for the rear seat. The F-16B prototype flew in August 1977.

Inevitably there were problems in the development phase -- instability at high alpha; glitches with the radar -- but these were easily overcome. More serious were troubles with the F100 turbofan which, given its history with the F-15, was only to be expected. In fact, because of the agility of the F-16, it was worked even harder, with more frequent excursions from flight idle to full augmentation, than it was in the bigger fighter.

The fact that the F-15 was chronologically ahead of the F-16 meant that many of the F100's problems had already been solved. This was the case with the "proximate splitter," a device which reduced the possibility of stagnation stalls, which was fitted only to the GD fighter.

SERVICE AND COMBAT
The first operational F-16 unit was the 388th TFW, at Hill AFB, which also acted as a conversion and training unit. From then on the build-up was rapid, and the Viper became a familiar sight in many countries around the world. The first overseas air force unit to equip with the type was the Belgian 349 Squadron.

By 1980, the lightweight fighter concept had vanished. Multi-role requirements had taken over, and the Viper became an all-purpose bomb truck, even with the USAF. The cry for ever more capability has since ensured that the upgrading process has been continuous.

As at spring 1999, the Viper has been credited with many air victories. The first was a Syrian Mi-8 helicopter, shot down with cannon fire by a young Israeli pilot on April 28 1981. The first fast jet victim, a MiG-21, fell to Israeli Amir Nahumi on July 14. In the Beka'a action of 1982, Nahumi accounted for six more Syrian aircraft to become the first, and so far the only, F-16 ace. The F-16 was credited with 44 air victories in this conflict.

Border violations during the Afghan War saw Pakistani Viper pilots shoot down at least 16 intruders. The Gulf War of 1991 was notable for a complete lack of opportunity for USAF F-16s, but on December 27 1992 a Viper notched up the first AIM-120 Amraam kill against an Iraqi MiG-25. On January 19 1993, an Iraqi MiG-23 was destroyed with the same weapon type. F-16s have shot down several Serbian aircraft between 1994 and 1999. Their victims include MiG-29s, one of which fell to a Dutch F-16. The Viper's air combat record to date is about 75 victories for no losses.

FURTHER DEVELOPMENT
The F-16 has had more metamorphoses, planned and actual, than any other modern fighter. A proposed forward-swept wing variant was never built, while the cranked tailless delta F-16XL and F-16F, stemming from interest in supercruise, both made it into

hardware, but were not adopted. Then came the YF-16CCV, the F-16 AFTI, and the F-16 VISTA, all of which were built to explore unconventional flight modes. Meanwhile some F-16As were modified for the air defense role, with improved APG-66 radars and AIM-7 Sparrows and AIM-120 Amraams.

The first major variant to enter service was the F-16C/D, however. This carried the Westinghouse APG-68 radar, with several new air-to-air modes and greatly improved detection and tracking ranges and capabilities. Sparrow and Amraam could be carried, the cockpit was upgraded with two MFDs and a wide-angle HUD, while a triplex digital FBW system was introduced without problems, according to Chief Test Pilot Steve Barter. Externally, the F-16C and D vary from the A/B only in the tail root housing, which was enlarged to house the ALQ-165 ECM gear.

Weight growth was inevitable, which degraded performance and maneuverability. To restore the original capability, two new programs were instituted. The first, for Block 50 and Block 52 aircraft, engines of greater thrust were proposed: the F110-GE-100 and the F100-P-229, respectively. But while these largely restored the original thrust/weight ratio, they did nothing to improve wing loading. Then in 1988, the Agile Falcon program, involving a new composite wing with a 25 percent increase in area and a commensurate increase in span, was proposed.

In the event, Agile Falcon was not adopted, it being too radical a change to accept retrofitting. The improved engine program was adopted, however, even though it does little more than make the existing wing work more inefficiently at high alpha.

Since then the F-16 VISTA has been fitted with a multi-axis vectoring nozzle, which allows it to explore the post-stall maneuver regime. But, since the F-16 is due to be replaced by the Joint Strike Fighter, it is doubtful whether this variant will ever enter service. Finally, there is the Block 60 F-16ES, with conformal fuel tanks above the wing/body blending.

TOP: Belgian Air Force No 349 Squadron was the first foreign unit to operate the F-16. This F-16A carries four Sidewinders in the counter-air mission.

RIGHT: A two-seater F-16B of the Republic of Korea Air Force hurtles skywards. The two tip-mounted AIM-9 Sidewinders were the sole missile armament originally envisaged.

USA

BOEING F/A-18 HORNET

The Hornet started life as the Northrop YF-17, the loser in the LWF competition against the YF-16, albeit by a smaller margin than most people realized. One factor had been that, whereas the YF-16 used the fairly mature F100, the Northrop contender was powered by two experimental General Electric YJ101 turbofans, which lacked the thrust that the developed engine would finally achieve.

When it came to the fly-off, the twin-engined configuration provided a built-in head-wind. Although the two YJ101s provided rather more thrust than the single F100 of the YF-16, this was offset by the extra structure and systems weight, and the volume and drag of a twin layout. The YF-17 was about 25 percent heavier than the YF-16.

Like General Dynamics, Northrop used computer-controlled variable-cambered wings, but with very large and subtly contoured root extensions. These not only provided a destabilizing effect to minimize the effect of trim drag at supersonic speeds but acted as compression wedges to reduce Mach number at the compressor face. This allowed the intakes to be set well aft, giving a short inlet duct.

This posed a design problem. The six-barrel M61A rotary cannon had been located in the wing roots on the F-14 and F-15, well away from the radar, which could be adversely affected by vibration when it was fired. But the contours of the YF-17 root extensions prevented a similar installation. Instead, it was sited high in the nose. The vibration caused when it fired did not affect the simple ranging radar fitted.

Unlike the YF-16, carbon-fiber composites were widely used for the skin of the YF-17: fuselage panels and access doors; wing leading and trailing edges; flaps; and vertical tail surfaces. Under the skin it was largely of aluminum construction. Like its GD rival, it used quadruplex analog fly-by-wire but, unlike the YF-16, it retained a mechanical backup "get you home" facility to the differentially-moving stabilizers, which unusually were more sharply swept than the wings, and of high aspect ratio. Twin fins and rudders were located forward, and were canted outwards to avoid them being blanketed at high alpha.

The cockpit was conventional, and avionics

ABOVE: From this angle the Hornet's large and carefully shaped leading edge root extensions are shown to advantage. Also visible on their upper surfaces are crude angles which modify the vortices.

F/A-18E SUPER HORNET

DIMENSIONS: Span 44ft 8½in/13.63m; Length 60ft 1½in/18.32m; Height 16ft 0in/4.88m; Wing Area 500sq.ft/46.45m²

ASPECT RATIO: 4.0

WEIGHTS: Empty 29,574lb/13,415kg; Takeoff c46,200lb/20,956kg

FUEL: Internal 14,400lb/ 6,532kg; Fraction 0.31

POWER: 2xF414-GE-400 augmented turbofans; Maximum Thrust 22,000lb/9,979kg; Military Thrust 14,000lb/6,350kg

LOADINGS: Thrust 0.95; Wing 92lb/sq.ft-451kg/m²

PERFORMANCE: V_{max} high Mach 1.8 plus; V_{max} low Mach 1.01; V_{min} c120kt/222kmh; Operational Ceiling 50,000ft/15,239m; Climb Rate n/a

WEAPONRY: 1x20mm M61A cannon with 570 rounds; Up to 12 AAMs, Amraam and Sidewinder.

USERS: (all Hornets). Australia, Canada, Finland, Kuwait, Malaysia, Spain, Switzerland, Thailand, USA.

ABOVE: This F/A-18C Hornet of VFA-82 carries AGM-88 Harm anti-radiation missiles in addition to two Sparrows and two Sidewinders.

were basic. The result was a superb close combat fighter, but it lost the ACF flyoff to a better, by a narrow margin.

CARRIER FIGHTER

Even as one door closed, so another opened. The Tomcat had proved unaffordable in the numbers required, and six more Phantom squadrons plus 18 fighter squadrons of the US Marine Corps were scheduled for re-equipment. In addition, 30 A-7 Corsair attack squadrons were scheduled to be replaced from 1980. What was needed was a dual-role fighter to replace both Phantoms and Corsairs.

Political considerations determined that the choice lay between the two ACF contenders. As neither GD nor Northrop had ever built a naval fighter, GD was teamed with Ling Temco Vought, and Northrop with McDonnell Douglas (more recently Boeing). After exhaustive evaluations the US Navy concluded that the YF-17 offered the greatest growth potential.

Carrier operations are arguably the most demanding missions of all. The airplane is catapulted off the deck, reaching 120kt (222kmh) in less than four seconds. Landing is even worse. Standard procedure is to open the throttles wide on touch-down to ensure a safe go-around if the wires are missed. When

the hook takes the wire, the airplane is brought to a halt in two seconds. In a worst case situation, a vertical speed of 24ft/sec (7.3m/sec) has to be withstood. Both take-off and landing make extreme demands on structural strength. Two other naval requirements are resistance to corrosion, and wing folding to optimize the limited space available in a carrier hangar - on the lift, or on deck.

Given these requirements, drastic weight increases were unavoidable. More power was needed, and General Electric scaled up the YJ101 by about 10 percent, with increased mass flow. With the minuscule bypass ratio of 0.20, the YJ101 had been unkindly dubbed "the leaky turbojet!" The bypass ratio was increased to 0.34, which improved specific fuel consumption, while the turbine entry temperature and pressure recovery ratio were increased. The result was the F404-400, with greatly improved performance.

Gradually McDonnell Douglas refined the YF-17 into the F-18, an altogether larger airplane, although it retained the basic lines of the original. Projected gross weight rose by 50 percent, with the prospect of more to come. Not the least of this was a 70 percent increase in internal fuel, needed in part to meet the more demanding endurance requirements of carrier operations. A flight refueling probe

RIGHT: The two-seater F/A-18F Super Hornet comes aboard the USS *Harry S Truman* during trials in March 1999. The complexity of the main gear is clearly visible from this angle.

was of course obligatory.

Greater wing area was needed to keep loading within acceptable limits; an extra 50sq ft (4.65m2) was added, partly by increasing the span by 2ft 6in (76.3cm), partly by extending the chord, while the root extensions were lengthened and refined to reduce approach speed and angle. The horizontal tail surfaces were also redesigned to reduce tail aspect ratio. In a weight reduction exercise, 40 percent of the surface was covered with advanced composites, more than on any other airplane then flying.

Whereas the flight control system of the YF-17 had been analog FBW, that of the new fighter was digital, making it the first production aircraft in the world to be so controlled. The mechanical back-up to the tailerons was retained, however.

The carriage of two AIM-7 Sparrows was a USN requirement, and the sides of the fuselage were reshaped to accommodate the big missiles semi-conformally on the corners. This brought problems in its wake. An extremely complex retraction system had to be adopted to allow the main gears to clear them, and this gave trouble for some considerable time.

ADVANCED TECHNOLOGY RADAR

The radar selected was the Hughes APG-65, the most advanced fighter radar of its time. What made APG-65 so outstanding was that it was the first production radar in the world to have a programmable signal processor. Previous radars were hard-wired, and had to be physically altered to take changes. With APG-65, modes could be modified, or new ones added, via the software. At the same time, Doppler filter and range gate functions were defined by program coding. This provided exceptional flexibility.

APG-65 had nine air-to-air modes, and many more for air-to-surface and navigation functions. It was in fact this capability that convinced the USN that both fighter and attack missions could be flown by a one-man crew. Prior to this there had been doubts.

Space precludes a full description of all nine air-to-air functions of the APG-65, but Range While Search can detect targets at all aspects and relative velocities out to about 80nm (148km), and Track While Scan is effective to about half this distance, keeping files on up to ten targets while displaying eight.

The other giant stride in the fighter world was the cockpit layout. McDonnell Douglas had pioneered HOTAS on the F-15. Now they set out to minimize pilot workload, which in a true dual role aircraft, promised to be horrendous. The solution was what came to be known as the "glass cockpit," in which virtually all the dials and instruments were replaced by three color MFDs. All flight, attack, navigation, weaponry and systems status could be called up at the touch of a

ABOVE: The F/A-18E Super Hornet differs from its predecessor in having trapezoidal intakes to its F414 turbofans; redesigned root extensions, and a dogtooth on the leading edge. It is to start entering service in 2001.

ABOVE: The Hornet has been very successful in the export market. This completely unarmed two-seater is flown by the Swiss Air Force.

button and, where necessary, shown on the HUD. One MFD was even linked to the RWR and ECM suite, and could diplay and identify threats. The F/A-18 cockpit set new standards for the fighter world.

DEVELOPMENT AND SERVICE

The first order was for 11 Full Scale Development aircraft, two of which were two-seaters. About 600lb (272kg) of fuel was sacrificed to make space. The first flight took place at St. Louis on November 18 1978, and trials continued apace. The fact that scaling up a proven design is not without its perils quickly became obvious. Higher than predicted drag caused shortfalls in range and acceleration; wing flexing under load made for a sluggish roll rate; and nosewheel lift-off speed was far too high.

A few tweaks, larger ailerons, and increased radius to the wing leading edge, corrected most of the problems. In the air, the Hornet, as it had been named, was easy to fly, had no alpha limits, and was almost spin-

proof.

The F404 was the greatest success story. It proved tolerant of disturbed airflow and throttle slams even at high alpha, was easy to airstart, and had a spool-up time of just four seconds. No serious problems were encountered.

The Hornet entered service with VFA-125 Rough Raiders in February 1981. This was a conversion training unit, and the first operational Hornet squadron was VMFA-314 Black Knights, from January 1983. The dual mission was causing problems, however; some pilots unofficially specialized in one of the roles without fully mastering the other. This was resolved by assigning primary tasks to squadrons – some to air combat, others to ground attacks.

One fault manifested itself only after prolonged service. At high alpha, vortices shed by the root extensions caused fatigue fractures in the tail and the rear engine mountings. The solution was to mount a short piece of angle on the root extensions to modify the vortices.

BETTER HORNETS

Like the F-16, the F/A-18 has been subjected to an almost continuous process of upgrading, although much of this was naturally concerned with the attack mission. The flexibility of the APG-65 was such that, software apart, it remained unchanged for a decade.

The first production F/A-18C was rolled out in September 1987, and the F/A-18D emerged shortly after. Most changes lay beneath the skin: greater capacity mission computers; improved ECM systems: provision for first six, then up to ten AIM-120 Amraam while retaining two Sidewinders on wingtip rails; ACES ejection seats; and many other things. Other, phased improvements continued. From 1992, weight growth made more power essential; more powerful -402 engines were installed in new aircraft. Then from May 1994 the radar APG-65 was superseded on the production lines by the APG-73, with tripled memory capacity and processing speed. Earlier machines are being upgraded with APG-73.

Hornets took an active part in the Gulf War of 1991 but, like the F-16 Vipers, were mainly engaged against ground targets. Their only air combat opportunities came on January 17, when two F/A-18Cs, loaded with air-to-ground ordnance, encountered two MiG-21s. A little switchology converted them to the fighter configuration; they downed both Iraqis with Sidewinders, then went on to complete their primary mission.

SUPER HORNET

When the stealthy Lockheed A-12 reconnaissance airraft was canceled in 1991, the F/A-18E Super Hornet was waiting in the wings. A development contract was awarded during the following year, and the first flight took place in November 1995.

While much of the avionics and systems were common to the F/A-18C, including APG-73, the Super Hornet is considerably larger and heavier. The wing area is 25 percent greater, and the planform varies only with the addition of a dogtooth on the leading edge to increase aileron authority. The area of the root extensions is 40 percent greater to allow alphas of 40 degrees or more. All tail surfaces have been radically increased in size, and the main gear simplified. Internal fuel has been increased by a massive 3,100lb (1,406kg). Weight has of course increased drastically, and new and much more powerful F414-GE-400 turbofans are used to restore the thrust/weight ratio. These have revised intakes of trapezoidal section, angled to

provide a modicum of stealth. While the design is not inherently stealthy, various measures are believed to reduce the RCS to about F-16 levels.

As at the spring of 1999, the Super Hornet is in low-rate production, and is expected to enter service in 2001, replacing the Tomcat and F/A-18A.

BELOW: An F/A-18D two-seater of VFA-125 Rough Raiders, a unit responsible for training pilots for the Pacific Fleet, and based at NAS Lemoore, California.

THE FUTURE

Fighter design is driven by the nature of the predicted threat. "Predicted" because the process is now so complex that the period between project initiation and service entry is generally something in excess of ten years. Then, when the aircraft finally becomes operational, it can be expected to remain in the front line for at least two - possibly four - decades. This places a heavy premium on getting the prediction right. If this is not done, fighter pilots must go to war with unsuitable equipment. But even if it is right, the long operational life places exceptional stress on reliability and maintainability, while adequate growth capability must be designed in from the start.

Fighters currently under development give an indication of predicted future trends, although an element of uncertainty exists. There is currently a strong accent on beyond visual range combat: detect first; identify first; shoot first. Although BVR combat has been with us for over 40 years, its record so far is not impressive. The main failing is in the field of positive identification. However, the possibility of massive improvements in this arena cannot be discounted. Reliable ID allied to longer-ranged AAMs with superior terminal homing and agility may make BVR combat the way to go.

Low observability, or stealth, plays an important part by significantly reducing detection range, which hopefully allows early detection and the first shot. As Pierre Sprey, former Pentagon fighter analyst, was fond of saying: "First shots count!" The advantages of the first shot cannot be overstated. Even if a kill is not achieved, those on the receiving end will become defensive. An enemy missile in the air concentrates the mind wonderfully, usually on survival at the expense of attacking options.

While stealth is widely regarded as a panacea, it must be treated with caution. There is not, nor is there ever likely to be, an undetectable fighter. There is always an outside chance that an advance in detection technology could render it obsolete overnight. Furthermore, once the fighter is within visual range of an opponent, most of the advantages evaporate. Large fighters are immediately disadvantaged against smaller opponents, as they are easier objects on which to maintain visual contact. For this reason, stealth cannot be allowed to compromise the traditional fighter virtues of performance and maneuverability.

One final point before we leave the BVR arena. Modern air warfare is very dependent on AWACS and tankers. If these can be destroyed from ultra-long ranges, much air power projection capability will inevitably be lost. While this is a big "if," it cannot be entirely discounted for the future.

Even when BVR combat proves largely successful, there can be no guarantee that the fight will not close to visual range; indeed the mission may demand it. Once within visual range, the primary requirements are agility and high off-boresight AAMs. We must not for one moment think that the latter reduce the need for agility. However maneuverable an AAM may be, a heart of the envelope shot still gives it the greatest chance of success. This in turn underlines the ability to bring the nose to bear rapidly.

Manufacturers' brochures notwithstanding, AAMs are not yet invincible. Theoretical probability of kill(Pk) is just that - theoretical - obtained in the white-coat, sterile, laboratory-type environment of the test range. The only true test is combat. Defensively, hard maneuver makes the fighter a far more difficult target for an incoming missile by making extreme demands on the AAM's guidance system, its maneuver capability, and even its fuzing, thus decreasing its Pk. Experience has shown that even a direct hit is not always lethal, and it does not take a rocket scientist to

ABOVE: The Lockheed F-22A Raptor, with its unparalleled combination of stealth, supercruise, agility, and advanced avionics, leads the world in fighter concepts. To preserve its stealthy qualities, all AAMs are carried internally.

LEFT: The Joint Strike Fighter (JSF) will supplement the barely affordable F-22A, and may also be acquired by Britain's Royal Navy to replace the Sea Harrier. Seen here is the Lockheed Martin contender in British colors.

RIGHT: Supermaneuverability, the ability to maneuver using vectored thrust rather than conventional aerodynamic surfaces and to retain agility at speed below the stall, is being explored. This is the F119-PW-100 with vectoring nozzle.

BELOW: The Rockwell/MBB X-31 was developed to explore post-stall maneuverability, using paddles in the efflux to vector the thrust. Like Typhoon, it had a long moment arm for the canard foreplanes. In mock combat it was very successful.

calculate that a miss within the so-called lethal distance is potentially even less so.

The next generation of fighters will be fitted with thrust vectoring. As has been demonstrated by experimental airplanes, this allows controlled flight outside the normal aerodynamic boundaries, and enables previously impossible maneuvers to be performed. It has other applications, stealth and short take-off among them.

AIRFRAMES AND ENGINES

In the mid-1970s, the F-16, with relaxed stability and fly-by-wire, marked a quantum leap in fighter maneuverability. At the same time it established a plateau which, despite the best efforts of fighter designers, proved very hard to exceed by any significant amount. Marginal improvements were possible, but at disproportionate cost.

Traditionally, the fighter performance and maneuver envelope are limited by the amount of lift generated for a given altitude/ speed combination, as related in the first section of this book. Altitude and maneuver therefore combine to shrink the flight performance envelope. It was obvious, however, that if some way could be found to maneuver effectively and under full control at speeds below V_{min}, tremendous advantages would accrue. In Germany, MBB's Dr Wolfgang Herbst formulated the concept of post-stall maneuverability.

This could not be done using conventional aerodynamic controls; something totally different was needed. The solution was thrust vectoring, making the thrust line of conventional fighters variable. Early experiments were carried out with simple paddles which extended into the efflux. So successful were these that all-axis movable nozzles were developed. These provide control at speeds where aerodynamic control surfaces -ailerons, stabilizers, rudders, canards, etc. - had lost their authority.

Vectored thrust is the future of fighter maneuverability. It makes redundant the old pilot saying, "Out of airspeed, altitude, and ideas!" Fighters equipped with it can not only maneuver well into the post-stall regime, but can perform previously impossible gyrations. To such a degree has vectored thrust encroached into the domain of control that it can no longer be separated from aerodynamics.

There is one major problem remaining, however. Unorthodox maneuvers, as notably demonstrated by the Rockwell X-31 at the Paris Air Show in 1995, are well within the province of test pilots. But how would the average squadron pilot fare? Some time previously the AFTI F-16 had experimented with unconventional maneuver - direct lift, side-force, etc, - but the probability of the average squadron pilot becoming disoriented was judged to be too high.

BELOW: Previously used to explore unconventional flight modes, the AFTI (Advanced Flight Technology Integrator) F-16 is seen here, minus intake-mounted fins, testing an automatic ground collision avoidance system.

it has been suggested that brain emissions could be connected to the flight control system, giving fly-by-thought. One's mind boggles at the effect that the standard fighter pilot expletive might have!

One problem currently being addressed is the effect of battle damage on the flight control system. Whereas a hit on one control surface or its actuator might make an airplane difficult or even impossible to control, this can be offset by a sensor system that reconfigures the FCS to compensate failures by using the remaining surfaces, including the flaps. Such a system was tested by the F/A-18E/F in late 1998, and its widespread use can be confidently forecast.

Human limitations are increasingly making the idea of an unmanned combat air vehicle (UCAV) attractive. There is no great difficulty in designing a fighter which can turn at 20g or more, far beyond the physical limits of a human pilot. But while this obviously gives combat advantages, what are they in real terms? Both radius and rate of turn operate on a law of diminishing returns. At normal - i.e., human pilot - combat speeds, typically 400kt (741kmh), the advantages are minimal. But in the high altitude supersonic regime, where turn radii are measured in miles rather than feet, the difference could easily become decisive.

UCAVs do offer many advantages. No longer would designers have to work to human physiological limitations. Having no pilot, no ejection seat, no FCS interface, and no environmental control system would reduce both weight and complexity. Stealth would be made easier: reducing the radar reflectivity of a cockpit is extremely difficult. (In this connection, radar reflections from the inside of the pilot's helmet are potentially quite large, even through his head!) The drag of the cockpit transparency would also be eliminated, improving performance.

UCAVs are already with us in a basic form, mainly for reconnaissance, while the Tomahawk cruise missile is in effect a "kamikaze" UCAV. The next step is a weapon-carrying recoverable machine, and studies are already in hand for the attack mission. But from this to the air superiority mission is a huge step. The question arises, is the UCAV to be completely autonomous, its mission fed into the software prior to take-off, or should it be remotely controlled from the ground?

Air superiority is a very complex field. To master this will take a long research period, and demand greater computer memory and processing speeds than are currently available. A problem here is basic fighter maneuvering. For each situation there is a countermove, but countermoves only play

Pilots have other limitations. About 6g is the maximum sustainable while flying offensively; 9g is the defensive limit. Another problem is g-induced Loss Of Consciousness (g-LOC). Measures have been taken to counter this, including a system which monitors brain emissions and takes control if the pilot is not fully alert. This is all very well in peacetime, but recovering the airplane into straight and level flight in the middle of a dogfight does not guarantee longevity.

Computer processing speeds and capacity can be increased, but the capacity of the pilot cannot. HOTAS and "glass cockpits" notwithstanding, a pilot can quickly become overloaded, unable to absorb one more piece of information. The search is on to reduce workload; a device called Pilot's Associate has been developed to assess the situation and advise the best action to take. Voice control is another option, while for the future the percentage game. In any given situation a countermove is something which works most, but not all, of the time. Air superiority will therefore be only as good as its program. The question whether our airplanes, systems, and weapons are better than those of the enemy then moves on to whether our programmers are better than the enemy programmers.

Another factor is all-round situational awareness. For BVR combat this has largely been overcome, but close combat is another matter. How critical is the loss of human flexibility, the ability to do the unorthodox when the situation demands? It will be a long while before artificial intelligence reaches this stage. The alternative is a man or woman in the loop. But this will make extreme demands on both detection and communications technology. One thing which must be replaced in the UCAV is the passive detection and identification sensor known as the human eyeball. Until such time as this can be done, the air superiority UCAV, whether autonomous or with a human controller, must remain a pipe dream.

LEFT: A whole gaggle of Lockheed Martin UCAVs, which appear to show that robot fighters can refuel in flight. However, in front is a piloted variant which may indicate that an airborne strike controller is needed to add flexibility.

BELOW: The Unmanned Combat Air Vehicle (UCAV), either controlled from the ground or preprogrammed, is currently under development to carry out attack missions. This stealthy design is by Lockheed Martin at Fort Worth.

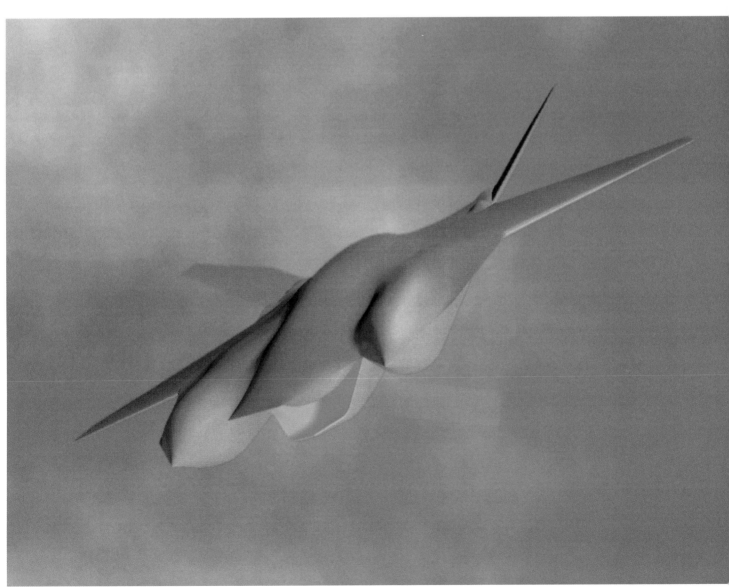

RIGHT: Stealth shaping is obvious with the Lockheed Martin JSF. The carefully angled intake lips, wing-body blending, canted fins, and chined radome obviously covering a phased array radar antenna, all combine to delay radar detection.

BELOW: Stealth features are also obvious on the Boeing contender for JSF, although this has a chin intake and a deeper fuselage. The wing leading edge shaping, and the cockpit canopy, which resembles that of the F-22A, are interesting.

STEALTH

Historically, the dominant factor in air combat has been surprise - the attainment of surprise on the attack, and the avoidance of surprise on the defensive. This is as true today as it was in 1914. Given this, the critical thing in air combat is to detect first and attack first. For all practical purposes this demands the avoidance of detection, by whatever means.

The time-honored means of detection is visual. This not only tells a fighter pilot that something nasty is lurking out there, it instantly tells him what it is doing, roughly how far away it is, and more often than not what it is. The disadvantages are short range, inability to penetrate darkness, cloud and haze, and - sadly - unreliability in identification. However, it is a passive system, giving no warning to an opponent that he has been seen.

Radar has certain advantages. It detects at far greater distances in darkness or adverse weather, and provides very accurate ranging and relative velocity information. On the other hand, it is an active system which betrays not only its presence but its location. It is also vulnerable to countermeasures.

Finally there is heat, or infra-red, detection. A passive detector, this works well in darkness, but is less good in cloud or rainy conditions. It has no basic ranging capability, but has excellent angular discrimination which can be used in conjunction with radar. Triangulation can sometimes be used to give range.

As radar is the primary air combat detection system, most stealth measures are directed towards defeating it. This is done by reducing radar cross section (RCS). For example, reducing the RCS to one tenth of its original value gives a reduction in detection range of about 44 percent. While this is enough to be tactically significant, much greater improvements are possible.

Physical size is only a factor in the visual arena; a small fighter is more difficult to spot than a larger one. It must be remembered that even under ideal conditions radar gives only a probability of detection, not a certainty. Stealth is therefore a means of reducing that probability by a combination of careful shaping and special materials.

The principle of shaping is simple. A ball hitting a wall at right angles bounces straight back to the thrower. At any other angle it bounces away. In the same way, shaping is used to deflect incoming radar emissions rather than reflecting them back to the sender. Of course, this can never be perfect; there will always be times when the aspect of the fighter to the radar will be such that the return is favorable. This is called a radar spike. By careful alignment of reflecting surfaces, the radar return can be concentrated into a few spikes, between which the return will vary between very weak and undetectable. The dynamics of flight will ensure that the time in which the spike is aligned will be very brief.

Stealth shaping consists of aligning all swept surfaces - wing, tail and fin leading edges and surface discontinuities such as control surfaces, undercarriage doors, etc. - at the same angle; the avoidance of vertical surfaces and right angles which act as corner reflectors (fins should be canted inwards or outwards, or where possible omitted altogether); and the masking of the engine compressor face by serpentine inlet ducts, even though this may reduce pressure recovery. In addition, an extremely precise surface finish is required, made more onerous by the need to sawtooth the leading edges of access panels. The number of these must be minimized, even though maintenance increases. Finally, external stores carriage must be avoided, whether weaponry or fuel tanks. But internal carriage demands a larger volume, and therefore a larger fighter. These are the major points; there are many minor ones.

The alternative to deflecting radar emissions is to absorb them. This is done by radar-absorbent material (RAM), which works by converting the electro-magnetic emissions into heat. This sounds worse than it is; the temperature rise is minimal. RAM comes in two forms: as rubber-based stick-on tiles (or ceramic-based in hot locations), or as paint. Tiles are used to line inlet ducts; coats of paint line wing leading edges, etc. Perhaps surprisingly, composites have little effect on stealth; if they are radar-transparent, metal structures inside them require shielding.

ABOVE: Another view of the Boeing JSF contender, showing internal weapons carriage. First projected as a tail-less delta, at the time of writing it is believed that a more conventional configuration is to be adopted.

Infra-red detection systems work at two main wavelength windows. The most IR-transparent atmospheric wavelength is between 8-13 microns (a micron is one millionth of a meter), making this an obvious choice, given the fact that IR is easily foiled by cloud, fog, etc. On the other hand, the largest and hottest part of any fighter is the engine efflux, most of the IR signature of which is caused by CO_2 in the exhaust gases. This is at

Radar Cross Sections

F-15 fighter	4,358 sq ft (405m²)
B-52 bomber	1,071 sq ft (99.5m²)
B-1A bomber	107.6 sq ft (10m²)
B-1B bomber	10.98 sq ft (1.02m²)
SR-71 reconnaissance	0.15 sq ft (0.014m²)
F-22A fighter	0.07 sq ft (0.0065m²)
F-117A recce/strike	0.03 sq ft (0.003m²)
B-2 bomber	0.02 sq ft (0.0014m²)

4.2 microns. So, for greatest effect, IR sensors work in two bandwidths: 3-5 microns as well as 8-13 microns.

Heat signatures cannot be disguised; they can only be lessened. For this reason, augmentation (afterburning) should be avoided where possible. At one time this would have been regarded as virtually impossible in combat, but with the advent of supercruise this is no longer the case. The greatest benefit of supercruise is to give the fighter plenty of energy for attack and disengagement without increasing its heat signature by augmentation.

Other methods of reducing IR signature are ejector nozzles, which mix ambient air with the efflux to promote cooling, while two-dimensional nozzles, as used on the F-117A, give the same effect by increasing the ratio of surface area to efflux, thus promoting mixing and cooling with ambient air. Yet another possibility is the use of a high-bypass ratio turbofan to reduce exhaust temperature, but a ratio in excess of 0.4 conflicts with the supercruise requirement.

So far we have discussed only the exhaust plume, but what of the engine itself? Curved efflux pipes can be used to shield the hot turbine section from direct view, although thrust losses must be accepted with this solution.

From any angle other than the rear quadrant, the main problem is kinetic heating of the airframe, basically to the nose and wing leading edges. Little can be done about this apart from keeping combat speeds well below Mach 2. In the air combat rather than the pure interception mission this should pose no real problems, while Mach 2 or more at disengagement will present the rear rather than the front of the airplane, in which case kinetic heating hardly matters.

While stealth has great potential value in combat, it is an extremely expensive commodity. A "gold-plated" stealth fighter is all but unaffordable, while stealth cannot be allowed to compromise traditional war-fighting virtues such as acceleration and maneuverability.

WEAPONRY

Like the ages of man in the Oedipus legend, air-to-air weapons come in three categories: BVR, visual, and last resort. The first two are AAMs; the latter the gun.

The current concentration on BVR combat makes this category of primary importance. Insofar as long range is concerned, the now elderly AIM-54 Phoenix missile is still hard to beat, but its weight and drag are unacceptable in the modern world. The current state of the art uses two-stage solid fuel rocket propulsion, with a booster, followed by a sustainer motor to give extended range. Typical V_{max} is about Mach 4, attained just a few seconds from launch, after which the AAM is coasting, gradually losing speed and energy. Mid-

RIGHT: While one might think that stealth has put the accent firmly on close combat, missile designers are seeking ever greater ranges. The Future Medium Range Air-to-Air Missile (Fraam) seen here, will use ramjets to increase its reach.

course updates from the parent fighter via data link keep it on track towards the target, and when the range closes to something under 10nm (18km) the active radar seeker cuts in to complete the interception.

This poses problems at extreme range. A radical change of course by the target at the wrong (right?) moment, may well take it outside the seeker look angle, in which case contact is lost. As speed decays, maneuverability reduces, making it difficult for the AAM to follow a hard-turning target. Even though targets may deploy ECM many AAMs have a "home on jam" capability, although this only works on certain types of jamming.

What of the future? The market leader in the West is the AIM-120 Amraam. The next step is the Future Medium Range AAM (Fmraam). This will have a rocket motor for initial launch, backed by a liquid-fueled ramjet. The extended burn time of the ramjet will not only extend operational range, it will ensure full maneuver capability out to current Amraam ranges. As a rule of thumb, the ability to engage a target maneuvering at 9g at 20nm (37km) allows a non-maneuvering target to be attacked at more than 30nm (56km). A ramjet-powered AAM would be far more lethal, expanding the "no-escape zone" by at least 250 percent.

One further point. Fmraam, and for that matter Asraam, can be cued by data from another fighter, from AWACs, or from the ground. This will allow attacks to be made without the parent airplane having to betray its presence by using its on-board radar, a tremendous advantage.

One attractive idea which now appears to have been laid to rest is a simple anti-radiation missile capable of homing on the emissions of fighter radars. In theory this is great; in practice the sophistication of modern fighter radars, with frequency agility coupled with the tendency to leave them on standby wherever possible while using targeting data from outside sources, makes it unworkable. As the saying goes, you could conquer the world with the weapon you haven't got! However, there is

ABOVE: The British answer to the agile Russian R-73 is Asraam, said to have much greater off-boresight capability, and outstanding speed and agility. Imaging infra-red using advanced software allows BVR shots, unusual in a heat-seeker.

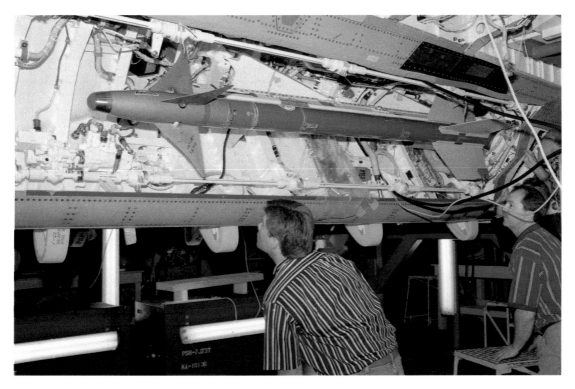

LEFT: Lockheed Martin engineers check an AIM-9M Sidewinder for size in the weapons bay of the F-22A. Similar checks have been carried out for Amraam. This is just one of the problems of internal weapons carriage.

ABOVE: An original solution to aerodynamic control surfaces on the Russian R-77 missile, widely known as Amraamski. Vympel state that far less power is needed to swivel these surfaces than to operate conventional fins.

a very attractive alternative: an ultra-long range AAM which homes on the emissions of AEW and AWACS aircraft, then switches to active radar or IR for the terminal homing phase.

Russian companies have been working on ramjet-powered projects since 1993. The Novator Ks-172 has a claimed range of 216nm (400km), while the Vympel R-77/RVV-AE has a reach of up to 70nm (130km). An interesting change is the "potato masher" rear control surfaces of the latter. In the close combat arena, the Vympel R-73 outclasses the latest variants of the Sidewinder, while helmet-mounted sighting has greatly increased off-boresight capability.

There has however been a great deal of rubbish written on the subject, not least that the R-73 can be used for an "over the shoulder" shot at a target in its rear hemisphere. Certainly the helmet sight can be used to cue a missile at targets in the rear hemisphere; equally certainly the R-73 can turn through more than 90 degrees. What must be remembered is that when launched, the R-73 has not yet acquired its target, and cannot do so until it has turned around to the point where the target comes into its field of view. Given firstly that it is a heat seeker, secondly that fighters fly in pairs or more for mutual protection, and thirdly that seekers are not yet clever enough to differentiate between red stars and white, there can be no guarantee that it will acquire the right target.

One thing is certain: the wingman of the firer, and anyone else in his formation, will become very upset if the leader tries an "over the shoulder" shot. Nothing is more guaranteed to break formation integrity. At the Farnborough Air Show in 1994, Hughes missile expert Ed Cobleigh assured the author that they had successfully developed an AAM with an off-boresight look angle of 120deg. But firstly it was extremely expensive; and secondly the probability of it homing on a friendly airplane was unacceptably high. Research concluded that it would take at least a decade to master this problem, and the project was dropped.

One solution might be Imaging Infra-Red (IIR). Whereas the first heat homers headed for the most prominent heat source, IIR allows a picture of the target to be built up. This gives IIR greater discrimination against IRCM flares; in the future it might well be able to provide an approximate identification of friendly, as opposed to enemy, aircraft. But this is still in the future.

The air weapon of last resort is the gun. This is for close range engagements only, typically for no more than 1,550ft (457m) range. In the days when the gun was the main air-to-air weapon, between six and 12 were carried. In modern times there is only one. Unlike most AAMs it is instantly available; a squeeze of the trigger unleashes a storm of fire at its opponent, and no countermeasures, other than hard maneuver, will avail against it.

Gun size is a compromise between caliber and muzzle velocity (i.e., hitting power), and rate of fire, which gives the greatest probability of hits. Experience has shown that the optimum for ballistic qualities is a shell of about 25/27mm, and most European cannon are in this range. The Europeans also tend to use the revolver mechanism which, while it gives the optimum between hits and hitting power, has a lower rate of fire than the multi-barreled M61A 20mm cannon favored by the USA.

AVIONICS

Fighter avionics are primarily concerned with three main areas: offensive detection, including identification and weapons usage; defensive detection; and countermeasures. All depend on computer processing capacity. A recent innovation in this field is the Gallium Arsenide microchip. GaAe chips not only offer up to ten times faster processing speeds than silicon, they require less power to operate, are effective over a wider temperature range, thus reducing cooling requirements, and are resistant to radiation, which enhances survivability. They therefore promise to be a major factor in the avionics field.

DETECTION

Radar is a primary means of offensive detection, but as an emission it is easily detectable by an adversary, which can usually gauge the degree of threat from the mode used, and take appropriate countermeasures - electronic or maneuver. For this reason, the most appropriate radar mode for many situations is standby, activating it at the last possible moment.

There is an alternative, however, which looks increasingly attractive. This is remote targeting, in which missile launch data is provided from off-board sources, such as other fighters, AWACS, or ground control. The air combat equivalent of the discovered check in chess, this allows surprise attacks to be made from unexpected quarters.

Remote targeting is of course very much a BVR capability, and in any case it is not always possible, in which case recourse to on-board radar is the only alternative. Then, once in visual range, the standard air combat radar modes come into their own.

ABOVE: Close-up of the Thomson-CSF RDY multi-role fire control radar. As can be seen, the slotted wave-guide antenna can be swivelled to quite a steep look-up angle. It can equally well be trained down or sideways.

BVR AND REMOTE TARGETING

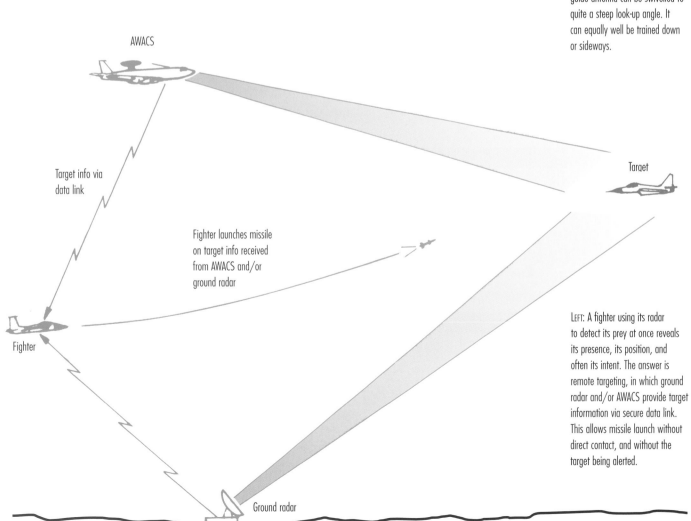

AWACS

Target info via data link

Fighter launches missile on target info received from AWACS and/or ground radar

Fighter

Target

Ground radar

LEFT: A fighter using its radar to detect its prey at once reveals its presence, its position, and often its intent. The answer is remote targeting, in which ground radar and/or AWACS provide target information via secure data link. This allows missile launch without direct contact, and without the target being alerted.

RADAR ANTENNA OPERATION

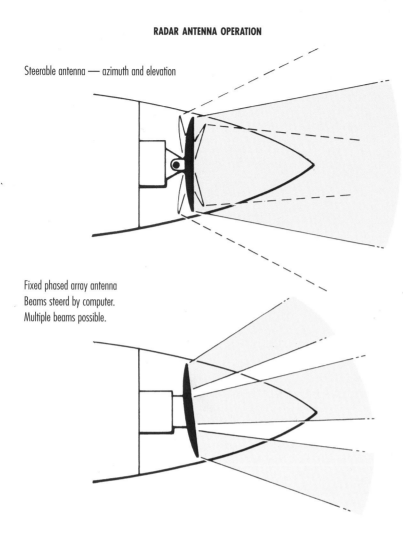

Steerable antenna — azimuth and elevation

Fixed phased array antenna
Beams steerd by computer.
Multiple beams possible.

ABOVE: Whereas the antennae on traditional radars have to be physically pointed to look in the desired direction, phased array antennae are fixed, and the beam is steered by computer. This makes multiple beams possible at once.

OPPOSITE: Simple countermeasures are often the most effective. Here a Tornado deploys chaff, metallic fiber cut to a specific length, to fool enemy radars, and flares to decoy heat missiles by confusing them as to the true target.

The traditional radar antenna is steerable, powered electrically or hydraulically to point at any desired area of sky within its angular travel. It has certain shortcomings, however. Its movement is relatively slow; the angle at which it is trained limits the volume of sky that can be scanned at any one time, and it is a good radar reflector - when it points directly at a target the antenna is in the ideal position to reflect enemy emissions straight back to the sender. Finally, the radome has to be large to accommodate antenna movement, and must be shaped for optimum transparency to its own emissions. Stealth is thus compromised.

The latest fighter radars use a fixed, phased array antenna which can be angled slightly upwards as a stealth measure. This has a spin-off effect in that it improves look-up capability. It consists of a battery of transmit/receive modules, and the beams (near-simultaneous mode operation) are steered electronically. There are many other advantages. The phased array gives a far greater ratio of listening to sending time, which increases the probability of target detection while decreasing the chance of its emissions being intercepted. Beam steering is faster and more accurate, while programming allows it to quickly back-

track to recheck a faint possible contact. This last gives it improved capability to detect stealth aircraft. Finally, not only can a much smaller radome be used, it can also be shaped for stealth. For the future, "smart skins," with send/receive modules built into the surface of the forward fuselage or wing leading edges, could replace the radar antenna altogether. But the technical difficulties of this are extreme, and the degree to which it would compromise stealth would appear to be excessive. Nevertheless, it remains a possibility.

However, there is an alternative to either active or remote radar detection. Radar Warning Receiver (RWR) systems have long been an essential part of the defensive avionics suite. At first they did little more than give visual and audio warnings that a hostile radar was painting the airplane and from which quarter.

Nowadays, greater computer capacity allows much more information to be extracted, and this can be put to good use. The hostile emission is compared with a "library" of enemy radar signatures, and positive identification is made; provided only that the fighter is not flying directly towards, or away from, the hostile radar, triangulation can give fairly accurate ranging information. Ranging data on the target can also be derived from other aircraft, or ground stations monitoring the same emission. Assuming that the source is an enemy fighter, bearing and range data thus obtained allow an AAM to be launched at it. This is of course far from easy since the frequency agility of modern fighter radars makes them difficult targets, but it is not impossible.

The other form of passive detection is of course IR. This was covered in some detail earlier in this book and little more needs be said about it except that, coupled with laser ranging, it can be used for tracking and targeting in close combat.

COUNTERMEASURES

It is not enough to know that something hostile is out there; measures must be taken to at least negate it. Electronic jamming is valuable but non-stealthy, while its emissions may betray more information than is entirely healthy. In any case, against modern frequency-agile, low sidelobe radars, it is becoming increasingly difficult.

Survival often depends on defeating an enemy missile, but the clever bit is knowing when it has been launched. Oxides in the missile exhaust plume can sometimes be detected by radar, but after burn-out this possibility vanishes. Current Missile Warning Systems (MWS) use passive ultra-violet sensors, but the future appears to lie with the so-called "two-color" IR sensor, which is less prone to false alarms.

Complete spherical coverage is needed, and work is afoot to provide this by using six "staring" - i.e., non-scanning - arrays, each with a 90deg x 90deg field of view, which would supply high-resolution imagery to the HUD. These must be conformal to meet stealth requirements. But these are warning systems only; thwarting a missile in flight is very difficult, made more so by the tremendous maneuverability of the latest breed.

Hard maneuver, ECM - either might work, but there is no guarantee. The next line of defense is the decoy. Air-launched decoys, their radar signatures enhanced to give the appearance of full-size airplanes, soaked up the Iraqi air defenses in 1991. Currently the accent is on towed decoys which, streamed astern of the parent fighter, use deception techniques to lure a radar missile onto themselves.

The advent of IIR missile guidance has made flares less effective, and other countermeasures are under consideration. Among these are lasers to confuse IIR seekers, although the level of accuracy needed is phenomenal. If it can be made to work, the obvious next step is missile-killing lasers. The alternative seems to be small and agile anti-missile missiles, carrying either ECM or

warheads. Given the current state of the art, the former approach appears the more promising.

Modern air combat promises an information explosion. Can the pilot handle it, or will this prove yet one more factor in a switch to UCAVs?

BELOW: Airborne targets must be small and cheap, and yet reliably simulate real aircraft by having their radar and heat signatures suitably enhanced. This led to their use as decoys to soak up Iraqi defenses in the Gulf War.

RUSSIA
SUKHOI S-37 BERKUT

ABOVE: A new forward-swept wing aircraft by the Sukhoi Design Bureau had long been rumored, and several planforms had been illustrated in the Western press with the designation S-32. All were inaccurate!

When Russian pilot Eugeny Frolov demonstrated the thrust vectoring Su-37 at Farnborough 1996, it seemed that subsonic fighter maneuverability was approaching the ultimate. Prior to this, it was rumored that Sukhoi was working on a forward-swept wing project designated S-32, but given the advantages of thrust vectoring this seemed superfluous. It was therefore a surprise to the West when Igor Voitintsev first flew the FSW S-37 Berkut (Golden Eagle) at Zhukovsky airfield on September 25 1997.

The advantages of forward sweep are five-fold: higher usable lift, lower supersonic drag, better low speed handling, reduced wing bending, and better area and volume distribution. This gives a smaller airplane for the same performance or better performance for the same sized airplane. But, whereas conventional aft sweep allowed the leading edge to twist downwards under high loadings, forward sweep twists the leading edge upwards, causing structural divergence. This builds so quickly that wings can be torn off before the pilot can take remedial action. Only the advent of aero-elastic tailoring using advanced composites made greater sweep angles possible.

It is obviously this that the S-37 seeks to exploit, and it is probable that it may couple FSW with thrust vectoring to provide close to the ultimate in combat maneuverability. FSW is of course far from new. Its benefits were first explored by the Grumman X-29A technology demonstrator between 1984 and 1991, but it was deemed not worth pursuing, and the project lapsed. Reasons have never been given, but it does not take a rocket scientist to see that while thrust vectoring represents a possible upgrade to existing fighters, FSW demands a radically new design.

Berkut is a development of the Su-27 Flanker, but varies considerably in detail. The wing planform is similar to that of the X-29A, with short inboard root extensions, and small close-coupled canard surfaces are carried ahead of it. Berkut is dimensionally rather larger than the standard Flanker, but the horizontal stabilizers are rather smaller. These are carried on booms outboard of the engines, each of which terminates in what appears to be a dielectric dome. The right-hand boom is rather longer than that on the left, and may contain an anti-spin parachute. The "sting" between the engine nozzles is absent, and the twin fins are canted outboard.

SUKHOI S-37 BERKUT

DIMENSIONS: Span 54ft 9½in/16.70m; Length 74ft 2in/22.60m; Height n/a; Wing Area n/a;

ASPECT RATIO n/a

WEIGHTS: Empty c43,692kg/19,819kg; Takeoff 52,960lb/24,000kg

POWER: 2xSaturn Al-31F augmented turbofans; Maximum Thrust 27,560lb/12,500kg; Military Thrust 16,755lb/ 7,600kg

FUEL: Internal 8,818lb/4,000kg; Fraction 0.17

LOADINGS: Thrust 1.04; Wing n/a

PERFORMANCE: V_{max} high Mach 1.60

WEAPONRY: None

USERS: None

At the front end, the huge two-dimensional inlets have been replaced by quarter-circular ones, apparently fixed, and similar to those of the F/A-18 Hornet. They are topped with a large chine which acts as a compression wedge. It has been stated that thrust vectoring nozzles will be added, but as at early 1999 there is no sign of this.

The Russians have not yet lost the habits of secrecy and confusion honed in the bad old days of the Cold War, and much uninformed speculation prevails in the West. The twin engines are stated to be Aviadvigatel D30-F6s as used on Foxhound, but as the S-37 is so closely based on the Su-27 it seems improbable that the Saturn Al-31F was not used.

Former Sukhoi Chief Test Pilot Major General Vladimir Sergeivitch Ilyushin, an old friend of the writer, has provided information on the project, amongst which was a maximum thrust figure which exactly matches the Saturn engine, but not the D30. Ilyushin also states that the S-37 is an experimental airplane, funded by Sukhoi, for research on flight maneuverability at high alpha and its operational systems and avionics. It is not a prototype fighter, and has no provision for weaponry. In addition, the fuel fraction is far too low for operational usage. Answers to the various anomalies in the Berkut are not forthcoming; we can speculate, however. The simplest answer to why the Berkut has not apparently been developed further is extreme financial constraints poervading the whole of Russia's services and general economy. It seems very probable that the S-37 is essentially a proof-of-concept machine, with Sukhoi deciding pragmatically that everything not actually needed for the flight test program should be omitted.

If this is actually the case, the Berkut is a full-scale fighter prototype but minus all inessentials. The low fuel fraction can be explained by the fact that internal fuel tankage and systems are minimized, allowing just sufficient capacity to enable the test program to be flown. In this case there will be no integral wing tanks; the omission of these will avoid potential problems, which need to be addressed only when (and if) the wing proves to have substantial advantages over more conventional forms.

By the same token, weapons stations and witing can be added as necessary, as can a "full-up" avionics suite. Doubtless all will be revealed in the fullness of time!

BELOW: This angle shows the forward-sweep planform of the S-37 to advantage, also the stubby close-coupled foreplanes. The inboard wing section has an orthodox forward-sweep chine.

USA

LOCKHEED MARTIN F-22 RAPTOR

ABOVE: Seen here when first rolled out, the YF-22 was less futuristic in appearance than the YF-23, but more workmanlike. Almost certainly it was less stealthy than its rival but had greater potential in close combat.

The origins of the F-22 lie in the Advanced Tactical Fighter (ATF) studies begun in 1981, aimed at defining a replacement for the F-15. The perceived threat was still the Soviet Union, and the requirement was not only to be able to operate offensively in the face of vastly superior numbers, but to counter the next generation of Soviet fighters. This was only partly a defensive requirement; wars are won by offensive rather than defensive action, by carrying the fight to the enemy.

This posed problems. Defensive fighter operations are carried out in a relatively benign environment, backed by ground radar, control and ECM, and SAMs. There is

less need for AWACS and tankers, and in any case these can be located well back out of harm's way. By contrast, offensive operations must be carried out in the teeth of the enemy air defense system; as the distance penetrated increases, the benefits of friendly ground assistance diminish, while missions have to be flown at a greater distance from the traditional force multipliers, flight refueling and AWACS. How best to do this? A compound solution was sought, based on the three-pronged technologies of speed, stealth, and agility.

Absolute speed, or V_{max}, was not really a player here, although a requirement for Mach 2 plus was unavoidable for the pure intercep-

LOCKHEED MARTIN F-22 RAPTOR

DIMENSIONS: Span 44ft 6in/13.56m; Length 62ft 1in/18.92m; Height 16ft 5in/5.00m; Wing Area 840sq.ft/78.04m²;

ASPECT RATIO 2.36

WEIGHTS: Empty 31,670lb/14,365kg; Takeoff 55,000lb/24,948kg

POWER: 2xF119-PW-100 T/V augmented turbofans; Maximum Thrust c35,000lb/15,876kg; Military Thrust c28,000lb/12,701kg

FUEL: Internal c25,000lb/11,340kg; Fraction 0.45

LOADINGS: Thrust 1.27; Wing 65lb/sq.ft-320kg/m²

PERFORMANCE: Vmax high Mach 2 plus; Vmax low Mach 1.21; Vcruise Mach 1.43; All else classified

WEAPONRY: 1x20mm M61A-2 cannon with 480 rounds; 6xAIM-120C Amraam

USERS: USA

tion mission. But for offensive sorties, when penetrating hostile territory, endurance dictated that baseline speeds were in the high subsonic region. Augmentation was used only when combat was joined; only rarely did speeds become supersonic, and even then fuel consumption became so high that it could not be held for long.

One thing was easily calculable, however. Speed was a major factor during offensive operations. As the SR-71 had conclusively demonstrated, SAM engagement envelopes could be crimped-in enormously by a combination of high speed and altitude. High speed also posed tremendous problems for opposing fighters. First, they had to attain a valid attacking position, and the higher the speed the less time they had available to do it. Like SAMs, AAMs also had their envelopes reduced by high speed.

Supersonic speed was not a great advantage, however, if it ran the fighter out of fuel quickly. The answer was supercruise, a fairly high supersonic speed in military power only. Calculations determined that the minimum should be Mach 1.4, and this was set as a requirement for the ATF.

Mach 1.4 at high altitude is 802kt (1,487km/h) in standard terms, but far more relevant is that it is 1,355ft/sec (413m/sec). To intercept something traveling this fast calls for extreme precision. Even assuming that the defender can get into position for a head-on attack, which is highly unlikely, the time available to detect, lock on, and launch a missile is marginal.

From the rear quarter the odds are far worse. The critical factor is the delta velocity - the difference between the attacker and the interceptor. As this increases, the difference between missile and gun range evens out until there is little to choose between them.

Supercruise is therefore a close to ideal way of clearing six o'clock; at least until much faster AAMs are produced. The other advantage is that military power gives a much

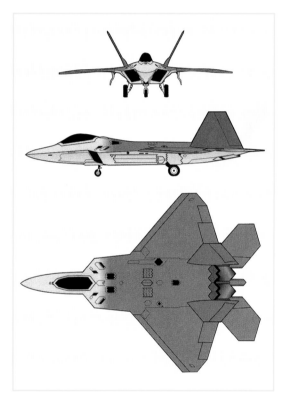

LEFT: Three-view of the F-22A Raptor, showing the revised wing planform, but not at this point the more orthodox fin and rudder shape. Its great advantage over the YF-23 was its ability to use thrust vectoring.

BELOW: Although the Northrop YF-23 was the losing contender in the Advanced Tactical Fighter contest, it shows many interesting stealth features, not the least of which are the trapezoidal wing planform and the flattened butterfly tail.

smaller heat footprint, while increasing operational radius. This was another critical requirement; an operational radius of 800nm (1,482km) was specified.

Stealth was an obvious requirement; it is difficult to fight the invisible man if one cannot see him. The RCS target for the ATF was set at one hundredth of that of the F-15, while emissions, radar, ECM, and communications, had to be tightly controled, and heat signatures minimized.

Other requirements were a short-field capability of 2,000ft (610m), and a clean take-off weight of 50,000lb (22,680kg). This last was unusual. Traditionally fighters carried external stores; to meet stealth requirements for the air superiority mission, the ATF had to tote all fuel and munitions internally.

Maneuverability requirements included 9g sustained turns at Mach 0.9 with 80 percent internal fuel; 6g turns at Mach 2.5 at 30,000ft (9,144m); and 2g turns at 50,000ft (15,239m); plus a minimum turn rate of 12deg/sec across the board. Acceleration from Mach 0.80 to Mach 1.80 was to be achieved in just 50 seconds. This was to be done using augmentation, of course.

Two teams were contracted to produce prototypes: Northrop (the most stealth-experienced company) coupled with McDonnell Douglas, who had already collaborated on the Hornet, built the YF-23. Lockheed were teamed with Boeing and General Dynamics to produce the YF-22. Engine manufacturers were General Electric and Pratt & Whitney, with the XF120 and XF119, rated at about 35,000lb (15,876kg) maximum and 28,000lb (12,701kg) military thrust, respectively.

The first ATF contender to appear was the Northrop/McDonnell Douglas YF-23. This was optimized for stealth combined with high performance. A cropped trapezoidal wing, coupled with steeply angled tail surfaces, which combined the functions of stabilizers and fins (called by Northrop ruddervators) gave what can only be described as a "slippery" planform.

By contrast, the Lockheed/Boeing/GD YF-22 was rather more orthodox in appearance. So much so that it rather resembled an F-15 extrapolated for stealth. It had a cropped trapezoidal wing like the YF-23, but with a more moderate forward sweep on the trailing edge, and steeply canted fins and rudders. It also featured two-dimensional thrust vectoring, something that had been omitted on the Northrop/MCAIR fighter as compromising stealth.

After an extended flyoff period, the YF-22 was declared the winner in April 1991. At the same time, the Pratt & Whitney YF119 was selected to power the new fighter. This was surprising, since the General Electric YF120 had given an optimum supercruise speed of Mach 1.58, significantly more than the Mach 1.43 of the YF119. While not officially stated, specific fuel consumption, giving greater operational radius, was probably the deciding factor.

On October 2 1991, an order was placed for nine standard production models, plus two two-seaters. Redesign was in hand, and this resulted in a shorter overall length and a modified wing planform, but with the same area, although wing root thickness was reduced, and twist and camber were modified. The canted fins and rudders were reduced in size, while the stabilators were reshaped. Other changes involved the forward fuselage, the cockpit and the engine intakes. The new fighter was initially named the "Lightning II," but was later dubbed the "Raptor" - a creature, not necessarily a bird, of prey.

The need to carry all fuel and stores internally makes for a rather large fighter, such that stealth measures are important. Initially a gun was to be specially designed for the Raptor, using case-telescoped ammunition. With this, the projectile is surrounded by propellant, which allows many more rounds to be stored in the same volume. However, this did not work properly, and the M61A-2 Vulcan cannon is installed in the starboard wing root. New 20mm ammunition, with improved ballistic qualities, is being developed for this. For the future, homing ammunition, able to follow the target in flight, has been proposed, but cost will probably kill this proposal.

Main armament will consist of six AIM-120C Amraam, carried in three internal bays. When selected, each missile will swing out on a trapeze for immediate launch, after which the bay will snap shut. Aerodynamic changes caused by the bay opening will be compensated automatically by the flight control system. This is far from new; similar systems

LEFT: This view shows the revised fin and wing planform shape. From this angle it is evident that the thrust vectoring is two-dimensional. This not only aids performance in pitch but by relieving aerodynamic forces it improves roll rate also.

were used with the Convair F-102 Delta Dagger and F-106 Delta Dart interceptors. For non-stealthy - i.e., defensive missions - or against a lo-tech adversary, eight more Amraam or Sidewinders can be carried underwing.

Radar is the Northrop/Grumman APG-77 with active array scanning, coupled with VHSIC computing. Cockpit information is provided on six color MFDs, and Pilot's Associate can advise on courses of action during combat. Current orders are for 339 Raptors to equip three US Fighter Wings. Initial operational capability is scheduled for 2005.

BELOW: Although dimensionally only slightly larger than the F-15, the Raptor carries almost double the amount of internal fuel, which is needed for extended supercruise. The only fighter with comparable internal tankage is the Russian Su-27, but this is is non-stealthy.

INTERNATIONAL: UK/USA
JOINT STRIKE FIGHTER

ABOVE: From head-on, the X-35 has many similarities to a miniature F-22, which is hardly surprising considering Lockheed's involvement in both projects. One variation is inward-canted inlets, a shape not seen since the F-105 Thunderchief.

The keynote of the Joint Strike Fighter - JSF as it is currently known - is afford-ability, which is to be achieved by a combination of small size and a huge production run. In turn, the latter will come from it being a true multi-role fighter, tailored to the needs of no fewer than four different services, with others currently interested. However, the scale of the possible economies make it an attractive proposition to service chiefs who have seen their inventories shrink as the costs of modern technology rise, while the politicians seek budget savings.

JSF originated in 1986, when a combined American and British study looked at ways of providing a supersonic replacement for the Sea Harrier/AV-8B. The obvious way, plenum chamber burning applied to a derivative of the original aircraft, posed extreme technical problems, not the least of which were ground erosion, and hot gas recirculation and ingestion. There was also the fact that the basic layout, with a centrally mounted engine with four nozzles spaced around it, was not really suitable for more than marginal supersonic flight.

The study explored Advanced Short Take-off Vertical Landing (ASTOVL) concepts, advanced vectored thrust, remote augmentation lift and ejector lift. All proved unsatisfactory for a variety of reasons, while the tandem fan system was far too heavy, too complex, and unproven.

In 1990, the Advanced Research Projects Agency (ARPA), commenced the X-32 Common Affordable Light Fighter (CALF) study, as a purely American initiative. This concluded that the best approach was a single (not bifurcated, as in the Harrier) vectoring nozzle, which allowed the engine to be located in the traditional aft position, augmented by a lift-fan set forward. The lift-fan could be driven either by a shaft, or by high energy exhaust gases from the main engine.

The advantages of this were that the airflow

from the lift-fan would be low-energy and low temperature, thus minimizing ground erosion and hot gas ingestion effects. Anyone who queries the seriousness of this should recall that the Yak-141 Freestyle, which uses two forward lift jets, had to make a long and orthodox rolling take-off when demonstrated at the Farnborough Air Show in 1992.

The disadvantages are weight and complexity; the fact that the lift-fan has to be covered to reduce drag in forward flight; far too much wasted fuselage volume; and less than inspiring STOL capability. Nevertheless, a supersonic STOVL airplane was wanted, and in 1993 a technology validation project, with Britain once again involved, was initiated. The demonstrator was to be single engined, with a maximum empty weight of 24,000lb (10,886kg).

Initially there were three main contenders, all consortia formed by major companies to pool resources and expertise. McDonnell Douglas teamed with Northrop Grumman and British Aerospace; Lockheed, prime contractors on the F-22, teamed with General Dynamics (later they merged to become Lockheed Martin); the third consortium was led by Boeing. Pratt & Whitney and General Electric were the main engine contractors; both worked closely with Rolls-Royce on lift-specific items. But meanwhile, under the influence of two of the potential operators, much of the emphasis moved to Conventional Take-Off and Landing (CTOL).

The USAF needed long-term replacements for the F-16 and A-10, a total acquisition of

2,036 JSFs, while the USN preferred a carrier-compatible version to replace some 300 F-14s and those A-6s not supplanted by the F/A-18E. Had it not been for the USMC, the demand for ASTOVL capability might have lapsed altogether, but its projected buy of 642 JSFs to replace both AV-8Bs and Hornets was too important to be ignored, whereas a mere 60 for Britain's Royal Navy would have had little influence. Since then, the USAF has indicated that it may take up to 200 ASTOVL JSF

ABOVE: With huge fleet carriers at its disposal, the US Navy does not need STOVL, but its aircraft will need to be stressed for catapult launches and arrested deck landings as evidenced by the tail hooks seen on this computer-generated image.

LEFT: Lockheed Martin's Joint Strike Fighter contender, the X-35 is seen here as it would appear in US Marine, US Navy, and Royal Navy service. Both the USMC and British RN variants will feature STOVL, with a shaft-driven lift fan forward and a vectoring nozzle aft.

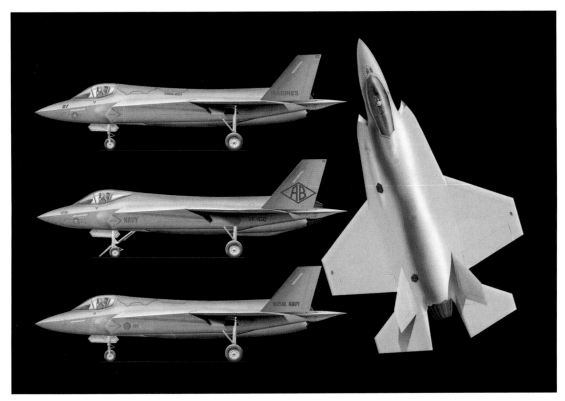

ABOVE: The Boeing JSF contender is the X-32 seen here in its original tail-less delta form, although it is now to have a swept wing and conventional tail surfaces. The large chin intake hinges downwards to increase the capture area for thrust-borne flight.

BELOW: The swivelling nozzle of the X-35 engine is a three-piece "lobsterback," as pioneered by Yakovlev for the Yak-141. There have been close design links with Yakovlev, and the Soyuz engine company has provided technical assistance.

variants to give greater flexibility during deployments to regions where airfield facilities are scarce.

This notwithstanding, it means that the standard CTOL JSF will be penalized by having a configuration compatible with ASTOVL when this capability is not needed. Other differences are flight refueling receptacles for USAF aircraft, and retractable probes for the others. Common to all is a high degree of stealth; a relatively high fuel fraction, probably 0.40; internal weapons carriage, with at least four underwing hardpoints; and a phased array radar.

Whether an integral gun will be carried is a moot point. The USMC prefers podded cannon, and the RN will probably follow suit, but the USAF and USN want an internal gun.

On the other hand, they are concerned about the volume and effectiveness of the 20mm M61 Vulcan, and studies are in progress for a larger caliber weapon, probably 25mm, which would use caseless ammunition with improved propellants. If successful, this would probably also be refitted to the F-22.

The McDonnell Douglas proposal was radical in that it used a flat-angle butterfly tail. The lift-fan concept had been abandoned in favor of a Rolls-Royce turbofan lift engine behind the cockpit, coupled with a clamshell to divert the main engine efflux to two rotating nozzles for take-off and landing. In wing-borne flight, the clamshell was opened and the efflux switched back through the aft nozzle. McDonnell Douglas was however eliminated from the competition in November 1996 on the grounds of high technical risk involved in converting a proposal into operational hardware. That left the Boeing and Lockheed Martin proposals, now designated X-32 and X-35 respectively, each of whom will produce two demonstrator aircraft.

BOEING X-32

Boeing has recently taken over McDonnell Douglas, whose expertise has no doubt contributed greatly. The X-32 has a cropped trapezoidal wing, with a shallow forward sweep on the trailing edge. It has no horizontal tail surfaces, but two steeply canted fins. Power is a modified F119, unusually located amidships, although the augmentor and a two-dimensional vectoring nozzle are in the traditional rear position. For the ASTOVL mode, the vectoring nozzle is closed, and the efflux is diverted to two rotating nozzles set beneath the center of gravity. To provide the extra thrust needed, the bypass ratio of the F119 has been increased to 0.60. This direct lift system makes the X-32 very similar to the Harrier in concept.

There are some unusual features. The chin inlet opens at low speed to improve pressure recovery in conventional flight. The rotating nozzles are enclosed to minimize drag. Because of the location of the nozzles, the main gears retract into wing housings.

Dimensions vary in detail. All variants are 45ft (13.72m) long, but spans differ. That of the STOVL version is 30ft (9.14m), which fits British carriers without recourse to wing folding; but the span of the others is 36ft (10.97m). CTOL and STOVL versions have empty weights of about 22,000lb (10,000kg), while the carrier fighter, with its beefed up structure and provision for deck operations, weighs in at some 1,984lb (900kg) heavier.

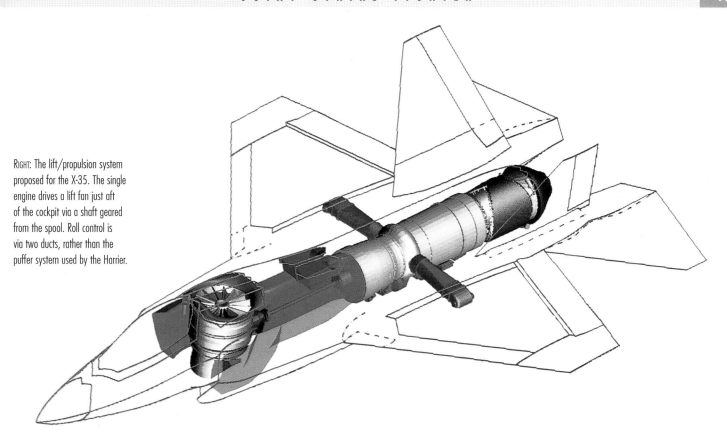

RIGHT: The lift/propulsion system proposed for the X-35. The single engine drives a lift fan just aft of the cockpit via a shaft geared from the spool. Roll control is via two ducts, rather than the puffer system used by the Harrier.

LOCKHEED MARTIN X-35

The Lockheed Martin group has since been joined by Northrop Grumman and British Aerospace, the latter having rejected offers from Boeing. Externally the X-35 closely resembles a small F-22 which, given its provenance, is hardly surprising. Power is again a modified F119 with a larger fan section. For the CTOL and carrier fighters, this will end in a low observable, axi-symmetric nozzle. But for STOVL, a three-piece vectoring nozzle will be used, similar to the "lobsterback" of the Yak-141. The other STOVL difference is a shaft-driven lift-fan just behind the cockpit. While Lockheed Martin have made light of it, engineering the multi-layer clutch and gearing must have been very difficult, given the rotation speed of a turbofan engine.

As with the X-32, the X-35 variants differ in detail. Length of all is 50ft 10in (15.50m), while span is 32ft 10in (10m), a limit set by the USMC. This means that the carrier variant needs outboard wing folding; it also has high-lift leading and trailing edges. Carrier approach requirements dictated a wing area of 540sq ft (50.2m²), while for the CTOL variant, USAF sustained maneuver specifications required an area of 450sq ft (41.81m²). Empty weight is predicted to be 21,600lb (9,798kg) for the CTOL variant, with an extra 1,400lb (635kg) for the extra STOVL kit. No figures are available for the carrier fighter.

No performance figures are available for either contender and, while JSF is almost certainly able to supercruise, Vmax is probably in the region of Mach 1.5-1.6. Agility requirements would at least match the F-16 in the subsonic and transonic regimes.

The demonstrators are scheduled to fly in 1999-2000. Whichever is the winner, deliveries are scheduled to commence in 2007, always assuming there is a winner, and that JSF is not canceled. The lethal political question is: where is the threat? But even if it is canceled, JSF has signposted the way ahead.

BELOW: In wingborne flight, doors enclose the lift fan of the X-35. The biggest potential customer is the USAF, which has indicated that it may acquire about 2,000 JSFs to replace the F-16 and A-10. About 100 of these may be STOVL-capable.

INDEX

Page numbers in **bold type** indicate references in captions to photographs and diagrams.

PICTURE CREDITS

Jacket: Front, Saab; back, Dassault. Endpapers: British Aerospace (BAe) via RB. Page 1: Lockheed Martin via M. Spick. 2-3: Artur Sarkisyan, Military Industrial Group "MAPO" via M. Spick. 4-5: Pratt & Whitney via RB. 7: Boeing. 8: Dassault Aviation via RB. 9: Top, Matra/BAe Dynamics, via RB; bottom, Lockheed Martin, via M. Spick. 10: Top, McDonnell Douglas, via M. Spick; bottom, Francois Robineau (Dassault/Aviaplans) via RB. 11: Top, Pratt & Whitney, via M. Spick; center, Pratt & Whitney via RB; bottom, BAe via RB. 12: General Dynamics via Salamander Picture Library. 13: Dassault Aviation via RB. 14: US Department of Defense via RB. 15: Francois Robineau (Dassault/Aviaplans) via RB. 18: Oliver Photographic. 19: Top, Oliver Photographic; bottom, M. Spick. 20: Oliver Photographic. 21: MATRA/CEV via M. Spick. 23: Top, MATRA BAe Dynamics via M. Spick; bottom, General Dynamics, via M. Spick. 24: Top, Raytheon Systems Company via RB; bottom, Oliver Photographic. 26: Top, Military Industrial Group "MAPO" via M. Spick; bottom, Lockheed/Boeing/General Dynamics, via M. Spick. 27: Saab via RB. 28: Top and bottom, Oliver Photographic. 29: Top, US Department of Defense via RB; bottom, BAe via RB. 30: McDonnell Douglas via M. Spick. 31: Boeing via RB. 33: Grumman via M. Spick. 35: Top, Ericsson via RB; bottom, Raytheon Systems Company via RB. 36: Francois Robineau (Dassault/Aviaplans) via RB. 37: Martin Baker via RB. 38: Top and bottom, Martin Baker via RB. 39: Top, McDonnell Douglas via M. Spick; bottom, M. Spick. 40: Top, TEKPHOT/Dassault via RB; bottom, Saab via RB. 41: US Air Force via M. Spick. 42: Oliver Photographic. 43: Top, Saab; bottom, Pratt & Whitney via RB. 44: BAe via RB. 45: Top, MATRA/BAe Dynamics via RB; bottom, Oliver Photographic. 46: Top and bottom, China National Aero-Technology and Export Corporation via M. Spick. 48: Oliver Photographic. 49: China National Aero-Technology Import and Export Corporation via TRH Pictures. 50: BAe. 52: BAe. 53: Geoff Lee/BAe via RB. 54: Eurofighter GmbH via RB. 56: Top, BAe; bottom, BAe via RB. 57: BAe via RB. 58: Rolls-Royce via RB. 59: BAE via RB. 60: Dassault Aviation via RB. 62: Francois Robineau (Dassault/Aviaplans) via RB. 63: Francois Robineau (Dassault /Aviaplans) via RB. 64: TEKPHOT via RB. 65: Francois Robineau (Dassault/Aviaplans) via RB. 66: Francois Robineau (Dassault/Aviaplans) via RB. 68: Top and bottom, Francois Robineau (Dassault/Aviaplans) via RB. 69: Alain Ernoult/Dassault Aviation via RB. 70: BAe. 72: McDonnell Douglas. 73: BAe. 74: Kevin Wills/TRH Pictures. 76: Top, E. Nevill/TRH Pictures; bottom, Mike Brandt/McDonnell Douglas via M. Spick. 77: Artur Sarkisyan, Military Industrial Group "MAPO" via M. Spick. 78: US Department of Defense via TRH Pictures. 80: Oliver Photographic. 81: G.D. Taylor/TRH Pictures. 82: G.D. Taylor/TRH Pictures. 84: Top, G.D. Taylor/TRH Pictures; bottom, E. Nevill/TRH Pictures. 85: E. Nevill/TRH Pictures. 86: Top, G.D. Taylor/TRH Pictures; bottom, E. Nevill/TRH Pictures. 87: E. Nevill/TRH Pictures. 88: Katsuhiko Tokunaga/Saab. 89: Enator Miltest AB/Saab. 90: Top, Katsuhiko Tokunaga/Saab; bottom, Ulf Fabianson/Saab. 91: Katsuhiko Tokunaga/Saab. 92: Oliver Photographic. 93: Oliver Photographic. 94: US Department of Defense via RB. 96: US Department of Defense via RB. 97: Top and bottom, US Department of Defense via RB. 98: Pratt & Whitney via RB. 100: Boeing. 1001: McDonnell Douglas via M. Spick. 102: US Department of Defense via RB. 103: Top and bottom, US Department of Defense via RB. 104: Katsuhiko Tokunaga/DACT via RB. 106: Northrop Grumman via RB. 107: Doug Moore via M. Spick. 108: US Department of Defense via RB. 109: Top, Lockheed Martin via RB; bottom, Oliver Photographic. 110: Boeing. 111: Oliver Photographic. 112: Oliver Photographic. 113: Boeing. 114: Boeing. 115: Boeing. 116: General Electric via M. Spick. 117: Top, Lockheed Martin via M. Spick; bottom, Lockheed Martin via RB. 118: Top, Pratt & Whitney via M. Spick; bottom, Deutsche Aerospace via M. Spick. 119: US Air Force Flight Test Center via M. Spick. 120: Oliver Photographic. 121: Top, Lockheed Martin via M. Spick; bottom, Oliver Photographic. 122: Top, Lockheed Martin via RB; bottom, Boeing via RB. 123: Boeing via RB. Hughes (UK) Ltd. via M. Spick. 125: Top, Matra/Bae Dynamics via RB; bottom, Lockheed Martin via M. Spick. 126: M. Spick. 127: Thomson-CSF via M. Spick. 129: Top, Northrop Grumman via M. Spick; bottom, Oliver Photographic. 130: Oliver Photographic. 131: Oliver Photographic. 132: Lockheed/Boeing/General Dynamics via RB. 133: Artwork and photo via Oliver Photographic. 134: Lockheed Martin via RB. 135: Lockheed Martin via RB. 136: Lockheed Martin via RB. 137: Top and bottom, Lockheed Martin via RB. 138: Top, Boeing; bottom, Lockheed Martin via M. Spick. 139: Top, Lockheed Martin via M. Spick; bottom, Lockheed Martin via RB.